Ecological Assessment of CROP BIOTECHNOLOGY

Ecological Assessment of CROP BIOTECHNOLOGY

Editor
Dharamvir Hota

2007

Gene-Tech Books

New Delhi - 110 002

ISBN 81-89729-17-9

Published by : GENE-TECH BOOKS
4762-63/23, Ansari Road, Darya Ganj,
NEW DELHI - 110 002
Phone: 41562849
e-mail: genetechbooks@yahoo.co.in

Printed at : Tarun Offset Printers
Delhi

PRINTED IN INDIA

PREFACE

Crop biotechnology offers significant improvements in virtually every area of crop production and utilisation, with potential benefits to farmers, the food industry, consumers and the environment. Transgenic crops, the main products of agricultural biotechnology, are increasingly becoming a dominant feature of the agricultural landscapes. The proponents of biotechnology argue that carefully planned introduction of these crops could reduce or even eliminate the enormous crop losses due to weeds, insect pests, and pathogens. They argue that the use of such crops will have added beneficial effects on the environment by significantly reducing the use of agrochemicals.

However, ecological theory predicts that as long as transgenic crops follow closely the pesticide paradigm prevalent in modern agriculture, such biotechnological products will do nothing but reinforce the pesticide treadmill in agroecosystems, thus legitimising the concerns that many environmentalists and some scientists have expressed regarding the possible environmental risks of genetically engineered organisms. In fact, there are several widely accepted environmental drawbacks associated with the rapid deployment and widespread commercialisation of such crops in large monocultures.

This book provides a comprehensive ecological assessment of crop biotechnology. The potential benefits and risks of genetically modified crops are addressed in detail. The book is an unique, concise and up-to-date resource, which offer students and researchers an innovative and valuable presentation of the subject.

Editor

Contents

1

Agricultural Evolution on Crop Biotechnology

Crops enhanced biotechnology are being grown on nearly 110 million acres in 13 countries. Food ingredients produced from biotech crops are found in thousands of food products consumed worldwide. However, while no unequivocal evidence of harm to our health or the environment from these crops is known or expected, there is an intense debate questioning their value and safety.

Genetically modified (GM) technology was first developed in the 1970s. Societal anxiety over this so-called GM food is understandable, and it is fueled by a variety of causes, including consumer unfamiliarity, lack of reliable information on the current safeguards in place, a steady stream of negative opinion in the news media, opposition by activist groups, growing mistrust of industry, and a general lack of awareness of how our food production system has evolved.

The scientific community has neither adequately addressed public concerns about GM foods nor effectively communicated the value of this technology. Clearly, societal acceptance is pivotal to the continued development and application of biotechnology in food and agriculture. Two decades ago, many agricultural scientists rightfully saw the emerging recombinant DNA technology as a potent tool in enhancing crop productivity and food quality while promoting sustainable agriculture. Much of this early excitement and expectation was met with successive breakthroughs in scientific research on plant gene transfer methods, identification of valuable genes, and the eventual performance of transgenic crops.

Plant breeders saw the technology as an additional means of crop improvement that could complement existing methods. For the first time, plant breeding was subjected to rigorous testing, and a regulatory framework was developed to oversee the commercialisation of GM crops on a case-by-case basis. There has been widespread acceptance and support for biotechnology from the scientific community. Accumulated experience and knowledge of decades of crop improvement combined with expert judgment, science-based reasoning and empirical research has led to scientists' confidence that GM crops may pose no new or heightened risks that could not be identified or mitigated, and that any unforeseen hazard will be negligible, manageable, or preventable.

Risks from GM crops should be monitored and measured, but concerns about these risks must also be balanced against the enormous benefits from this technology and weighed against alternative options. The strong trust that the American public has in its regulatory agencies (FDA, USDA, and EPA) has helped gain higher public acceptance of GM food in this country than in other nations. There are also larger and even more important sociopolitical issues such as anxiety about the control of food and agricultural systems, including questions about the pervasive impact of globalisation. Road thousands of years ago when plants were first domesticated.

Crop Evolution and Human Civilisation

Agriculture evolved independently in many places on this earth. For much of the 200,000 or so years prior to agriculture, humans lived as nomadic hunters, gatherers, and scavengers surviving solely on wild plants and animals. Subsequent domestication of these wild plants and animals from their natural habitats launched agriculture, thus radically transforming human societies.

This occurred initially in the Fertile Crescent, the Andean region in South America, Mexico, and parts of Asia, but diffused throughout much of the globe. A change from the nomadic lifestyle to farming led us to become community dwellers, eventually spawning the development of languages, literature, science, and technology as people were freed from the continuous daily task of finding food. Some regions caught on much faster than others, by margins of thousands of years. Plants have also evolved or, more accurately, they have been changed rapidly by human intervention.

Every crop plant grown today is related to a wild species occurring naturally in its center of origin, and progenitors of many of our crops are still found in the wild. Early humans must have tried eating thousands of feral plant species from a pool of a quarter of a million flowering plants before settling down on less than one thousand such species, which were subsequently tamed and adapted to farming. A little over 100 crop species are now grown intensively around the world, with only a handful of them supplying us with most of what we now eat. Through a process of gradual selection, our ancestors chose a very tiny section of the wild plant community and transformed it into cultivated crops.

Some profound alterations in the plant phenotype occurred during such selection, and these include determinate growth habit; elimination of grain shattering; synchronous ripening; shorter maturity; reduction of bitterness and harmful toxins; reduced seed dispersal, sprouting and dormancy; greater productivity, including bigger seed or fruit size; and even an elimination of seeds, such as in banana. These changes reduced the survivability of crops in the wild, and thus a feature that transcends all of our crops is the reduction of weedy traits from wild plants. Present crops are thus totally dependent upon human care for their survival, and modern crop varieties would persist in the wild "no longer than a Chihuahua would last in the company of wolves".

Most crops that supply our food were thus obtained at the end of the Stone Age, often from a relatively narrow pool of extant wild genetic diversity. Additional diversity arose within such cultivated crops through new mutations and natural hybridisation, and through judicious selection and perpetuation by farmers who maintained them as land races. Varied uses and preferences brought forth further diversification such as in corn (popcorn, sweet corn, dent corn, broom corn, and flour corn for tortilla and corn bread) or the derivatives of ancestral cabbage (kale, kohl rabi, brussels sprouts, cabbage, cauliflower, and broccoli).

With the advent of transoceanic navigation and the "discovery" of the New World, crops were moved around the world rapidly, often achieving prominence in adopted homes far beyond their natural centers of origin or domestication. For instance, the United States is the leading producer of corn and soybean in the word, yet these crops are native to Mexico and China, respectively. The world's largest traded commodity, coffee, had a humble origin in Ethiopia, but now much of it is produced in Latin America and Asia. Florida oranges have their roots in India, while sugarcane arose in Papua New Guinea.

Food crops that are now so integral to the culture or diet in the Old World, such as the potato in Europe, chili pepper in India, cassava in Africa, and sweet potato in Japan, were introduced from South America. Chinese gooseberry occurs wildly in China and is not edible. But careful breeding made it palatable, and it was re-christened "Kiwi fruit" in New Zealand after its introduction there early in the 20th century. The modern strawberry with big fruits is a product of the accidental crossing of two wild species from Virginia (United States) and Chile in France in the mid-18th century. Rapeseed, grown in India for centuries, was altered recently through classical breeding to eliminate the toxic erucic acid and smelly glucosinolates to result in canola-Canadian oil.

Triticale, a completely new crop, was artificially sythesised a few decades ago by combining the genomes of wheat and rye (two distinct genera that do not interbreed in nature). It is now grown on over three million acres worldwide. Modern bread wheat itself is also a fairly recent crop in the evolutionary time scale, having arisen only about 4,000 years ago through hybridisation of tetraploid (pasta or durum) wheat with inedible goat grass.

Mendel Evolution

While humans have always molded the evolution of crop plants, such changes imposed by farmers occurred over several millennia, leading to rich crop diversity—especially in traits related to their planting or consumption. At the same time, global population grew very slowly until the mid-19th century. It took 1,800 years for the global population to climb from an estimated 300 million around the time when Christianity began, to reach its first billion. But it took only 12 years to add the last billion, rising from five billion people in 1987 to six billion two years ago.

Fortunately, parallel scientific developments in agriculture ensured that food production kept pace with the population explosion of the past century. Beginning with Mendel's study of peas, knowledge of genetics helped usher in scientific crop development, resulting in high-yielding varieties. Food production increased in every part of the world in the past few decades, including in Africa. Per capita food consumption has also increased steadily everywhere except in parts of sub-Saharan Africa. In the United States and Canada, where such scientific developments and their applications were most intense, one average farmer now produces enough to feed nearly 150 people.

In crops subject to intensive scientific attention—corn, wheat, and rice—the productivity levels increased several fold. For example, U.S. corn growers averaged 26 bushels of corn per acre in 1928 and 134 bushels per acre in 1998. Such a prodigious increase in agricultural production was underpinned by scientific crop improvement methods along with other developments, including the use of irrigation, improved soil fertility management, mechanisation, and control of diseases and pests. To develop better crop varieties, scientists have used an array of tools. Artificial crossing, or hybridisation, helped us assimilate desirable traits from several varieties into elite cultivars.

When desired characteristics were unavailable in the cultivated plants, genes were liberally borrowed from wild relatives and introduced into crop plants. When a crop variety refused to mate with the wild species, various tricks were employed to force them to intermingle, such as the use of the carcinogenic chemical colchicine or by rescuing the hybrid embryos with tissue culture methods. Hybrid vigour was exploited in crops such as corn and cotton to boost productivity. When existing genetic variation within the cultivated germplasm was not adequate, breeders created new variants using ionising irradiation (gamma ray, x-ray, neutron), mutagenic chemicals (ethyl methane sulfate, mustard gas), or through somaclonal variation (cell culture).

Most people who are concerned about modern biotechnology have little or no knowledge of the processes that have been used to transform crops in the past. Nor are they likely aware that crops have been continually altered over time or that, without human care, they would cease to exist. Using a variety of tools over the past few decades, plant breeders have radically transformed our crop plants by altering their architecture (such as the development of dwarf wheat and rice), shortening growing seasons, developing greater resistance to diseases and pests (all crops), and developing bigger seeds and fruits.

These crops are also more responsive to management and better adapted to diverse ecological conditions. Improved food quality also resulted through fewer toxins (canola), better digestibility (beans), increased nutrition (high-protein corn), better taste, longer shelf life (thus withstanding long transportation and storage), and enhanced freshness in many vegetables and fruits. A 1,000-fold increase in the marble-sized wild Lycopersicon resulted in the modern tomato that can now weigh as much as a kilogram. Modern farming has steadily increased the supply of relatively safe, affordable, and abundant food not only in the developed world, but also in most developing countries.

An average American family now spends only 11% of its income on food and yet has access to better food choices with more variety and nutrition than ever before. Without scientific developments in agriculture, we would otherwise be farming on every square inch of arable land to produce the same amount of food. Using gene transfer techniques to develop GM crops thus can be seen as a logical extension of the continuum of devices we have used to amend our crop plants for millennia. When compared to the gross genetic alterations using wide-species hybridisation or the use of mutagenic irradiation, direct introduction of one or a few genes into crops results in subtle and less disruptive changes that are relatively specific and predictable.

The process is also clearly more expeditious, as the development of new cultivars by classical breeding typically takes from 10 to 15 years. The primary attraction of the gene transfer methods to the plant breeder, however, is the opportunity to tap into a wide gene pool to borrow traits, obviating the constraints of cross-compatible crop species.

With GM crops, at least we know what new genetic material is being introduced, so we can test for predictable and even many unpredictable effects. Consider, for example, how conventional plant breeders would develop a disease-resistant tomato. They would introduce chromosome fragments from its wild relative to add a gene for disease resistance. In the process, hundreds of unknown and unwanted genes would also be introduced, with the risk that some of them could encode toxins or allergens, armaments that wild plants deploy to survive.

To address the concern about long-term health consequences of GM foods, it is instructive to recognise that we worried little about such impacts when massive amounts of new proteins (and unfamiliar chemicals) were introduced into our foods from wild species or when unknown changes were created through mutation breeding. When new foods from exotic crops are introduced, we often assimilate them easily into our diets. Many so-called functional foods, health foods, and nutraceuticals have been entering into the mainstream American diet lately, with little or no regulation or testing.

In contrast, the GM foods currently on the market have been tested extensively and judged to be substantially equivalent to their conventional counterparts, with just one or two additional proteins present in miniscule amounts (introduced into a background of thousands of proteins). And, those proteins are broken down either during processing or digestion, with little long-term consequence. In food products such as oil, starch, and sugar, such proteins are not even found. A nagging potential problem with a new protein in food is that it could be a potential allergen.

There is no such thing as safe food, and there never has been. That is not to suggest that all of our foods are dangerous, only an acknowledgment that trace levels of such contaminants as toxins and carcinogens are present in everything we eat. But a primary rule of toxicology, articulated over 400 years ago by Paracelsus, refers to the importance of dosage: "Every substance is a poison, but it is the dosage that makes it poisonous". While not alarming, our daily food naturally contains thousands of chemicals, and many of them are shown to be carcinogenic or hazardous in lab animal studies with huge doses.

Here consume roughly 5,000 to 10,000 natural toxins daily, as plants have evolved to produce an array of chemicals to protect themselves against pests, diseases, and herbivores. For instance, roasted coffee has over 1,000 chemicals, of which 27 have been tested and 19 of them found to be rodent carcinogens. The fat-soluble neurotoxins solanine and chaconine are present in potatoes and can be detected in the bloodstream of all potato eaters. Naturally then, when crops are bred for resistance to pests by transferring genes through conventional methods, the resistance is often accompanied by an increase in such toxic compounds.

Thus, it is not true that we never had problems with conventionally bred varieties. Any crop variety found to pose a real health risk was promptly removed from the market, but those varieties (in contrast to GM crops) were never routinely tested. One pest-resistant celery variety produced rashes in agricultural workers and subsequently was found to contain 6,200 ppb of carcinogenic psoralens compared to 800 ppb in the control celery. This celery was removed from cultivation and that was also the case with the potato variety Lenape, which contained very high levels of toxic solanine.

Similarly, crop improvement practices evolved over time with continued refinement. It is common, though, for human nature to generate an exaggerated fear of new innovations while perceiving older or "natural" products as always more benign. This double standard in the larger context of risk regulation.

While most food hazards arise from pathogens such as Escherichia coli 0:157, Listeria monocytogenes, and Salmonella enterica along with mycotoxins produced by fungi (and thus a function of food storage and handling), certain foods containing toxic compounds are known to produce adverse health consequences over time. Cassava, eaten by a large population in Africa, contains cyanogenic glucosides, which cause limb paralysis if consumed before extensive processing. Solanin in tomato and potato is known to cause spina bifida. Vetch pea, a common legume known for its hardiness—and thus popular in India among poor farmers—contains highly dangerous neurotoxins that cause untold misery.

Phytohemagglutinin, found in undercooked kidney beans, is toxic. And peach seeds are extremely rich in cyanogenic glucosides. None of these were subject to any mandatory testing before they were introduced into the food chain, nor are they subject to any regulation now. But if the current regulatory standards imposed on GM crops were to be invoked for traditional crops, most of them would fail to meet their requirements.

Humans have built-in natural defenses that protect us against normal exposure to toxins. But, Toxic harmony with everything we eat, because natural selection occurs much too slowly and because much of what is in our diet today was not eaten at all when we were hunter-gatherers. A balanced mixture of foods normally provides adequate nutrition. However, none of the crops grown today were selected with our nutritional requirements in mind. Instead they were chosen intuitively, by our ancestors, from among the edibles that could be found around them. Thus, the most important food crop in the developing world-rice-has no provitamin A and little iron in its endosperm.

This has led to horrific problems, such as blindness among millions of children due to vitamin A deficiency, and iron-deficiency anemia in nearly a billion women dependent on a rice diet. Biotechnology research, far from causing any new food safety problems, has already demonstrated its potential in enhancing the nutritional quality of our food and is also being employed to reduce harmful toxic compounds that exist in our food.

Ecologically Demanding Food Production

All of us have to eat to live, and organised food production is the most ecologically demanding endeavour we have pursued. Agricultural expansion over the millennia has destroyed millions of acres of forest land around the world. Alien plant species have been introduced into non-native environments to provide food, feed, fiber, and timber, and as a result have disrupted local fauna and flora. Certain aspects of modern farming have had a negative impact on the biodiversity of crop plants and on air, soil, and water quality; nevertheless, it sustains and nurtures most of the world's six billion people with adequate nutrition and affordable food.

Here continuously introduced genes for disease and pest resistance through conventional breeding into all of our crops. Traits, such as stress tolerance and herbicide resistance, have also been introduced in some crops, and the growth habits of every crop have been altered. The risk of crop gene flow to weedy relatives has always existed, and such "gene flow" occurs where possible. Thus, it is comforting to recognise that no major "superweeds" have developed since the advent of modern plant breeding, although there have been a few instances of crops ever becoming weedy or of weeds becoming more invasive due to gene transfer from crops.

Most noxious weeds, such as kudzu, water hyacinth, and parthenium, resulted from the introduction of semidomesticated wild plants into non-native environments without the checks and balances of their native pests. Yet, there are probably no dwarf plants among the wild Oryza spp. and Triticum spp. populations in the Middle East or Asia, despite the fact that we now have been growing diminutive rice and wheat varieties for decades. The risk of gene transfer to wild plants is exacerbated when crops are planted in an area with compatible weedy relatives (as often seen in their centers of origin), when such species are promiscuous outcrossers (canola), or, most importantly, when the introduced genes enhance the reproductive fitness of the recipient weeds (although most genes introduced into crop plants, conventional or biotech, have little value in the wild).

The risk of gene transfer to weeds is similar with both conventional and GM crops and is not contingent on how introduced these genes into plants. Weeds do not become noxious as a result of any new crop variety. The current case-by-case testing and monitoring approach with biotech crops is a good regimen for the future, while the past experience with conventional crops provides assurance that such risks will be minimal and manageable.

Crop biodiversity is another issue of concern. The popularity of high-yielding varieties has already narrowed the genetic variation found in major crops. Biotechnology, if employed strategically, can reverse this through the recovery of older varieties that were discarded for lack of certain features (such as resistance to new disease strains), because modern gene transfer can restore such traits. Biotechnology research is also enabling the development of better methods for ex situ preservation of germplasm, such as cryopreservation, whereby valuable germplasm is being stored and thus saved from extinction.

The introduction of corn with a single transferred Bt gene has led to some concern about its ecological impact. While this concern should not be dismissed, it should be balanced with our hindsight and experience with corn itself, an introduced alien species now grown on 75 million acres in the United States, where none existed about 1,000 years ago. A crop introduced into a new environment entails the wholesale introduction of thousands of new genes. When grown on massive amounts of land, it exerts considerable ecological impact on the native fauna and flora, including beneficial insects. In contrast, the introduction of one or two genes into this background of 50,000 genes present in corn will have relatively less effect on the environment.

While the initial fear about the reported damage to monarch butterflies from Bt corn has not held up in additional studies, one also needs to consider the negative impact of alternate practices (such as pesticide sprays) and recognise the potential for positive impacts on beneficial insects by the GM crop due to the specificity of the insect target(s).

For that matter, any concern about "gene pollution" pales in comparison to the massive "risk" of alien crop introduction, as 95% of the crop area in the United States now consists of such introduced crops. Concern about horizontal transfer of genes from GM crops to other organisms, such as bacteria, has also been expressed. But it appears highly unlikely that the risk is dependent upon the method of gene introduction. An inherent feature of biotechnology is that it lends itself easily to molecular detection of introduced genes, but a true measure of risk can only come in comparisons with classically bred crops where little or no such studies have been performed.

Concerns such as random gene insertion, gene instability, and genomic disruption due to gene transfer have been expressed, but they are unlikely to be unique to GM crops or of any significance considering our current knowledge of genomic flux in plants. Worries about mixing genes from unrelated species ignore the history of plant breeding and the existing overwhelming sequence similarity of genes across kingdoms. Nevertheless, scientific research aimed at risk analysis, prediction, and prevention, combined with adequate monitoring and stewardship, must continue so that negative ecological impact from GM crops will be kept to a minimum.

Most problems raised by science can be solved by additional science itself. For example, appropriate promoters may ensure that pollen will not express genes toxic to beneficial insects, while gene expression strategies, such as sterile pollen, could reduce the risk of gene flow. One must also recognise the potential positive impact of GM crops on the environment, such as decreasing agricultural expansion to preserve wild ecosystems; improving air, soil, and water quality by promoting reduced tillage, reducing chemical and fuel use; improving biodiversity through resuscitation of older varieties and promotion of beneficial insects; and cleaning up contaminated soil and air through phytoremediation.

More exciting developments in biotechnology, such as genomics, and grapple with issues arising from consumer acceptance of innovations, historical knowledge on societal adoption of technological innovations may provide some valuable perspectives to scientists. Many innovations that would be good candidates for generating consumer apprehension and concern today were introduced in the past without concern because the public was less informed about innovation. The precautionary principle was never invoked to ensure the scientific certainty that crop varieties developed using nuclear irradiation or chemical mutagens were safe. And food labelling was never demanded for bread wheat improved with the addition of hundreds of unknown goat grass genes.

Many other innovations that are now commonplace in our lives were met with skepticism and opposition when first introduced. Such fear of technology was especially more pronounced in food related innovations. However, once consumers recognise that new innovations can enhance their quality of life and once they understood that risks are either minimal or manageable, such technology eventually could enjoy public acceptance.

This includes even those "disruptive" technologies that replace older ones. Nevertheless, there are historical instances of useful innovations that have not been readily accepted due to a variety of reasons, such as recalcitrance to adapt, entrenched economic interests opposing change, ideological opposition, exaggerated notions of risk (e.g. food irradiation), ill-timed product introductions, and serious conflicts with societal values and beliefs. Humans and crops will always be mutually dependent on each other's survival, and the guided evolution of crops will continue but increasingly will be more knowledge-based and responsible.

Benefits of GM Technology

Today's biotechnology is largely identified with applications in medicine and agriculture based on our knowledge of the genetic code of life. Various terms have been used to describe this form of biotechnology including genetic engineering, genetic transformation, transgenic technology, recombinant DNA technology, and genetic modification technology. One of the most prominent developments, apart from the medical applications, has been the development of novel transgenic crop plant varieties. Many millions of hectares of commercially produced transgenic crops such as soybean, cotton, tobacco, potato and maize have been grown annually in a number of countries including the USA, Canada (4 million), China (0.3

million), and Argentina (6.7 million). However, there has been much debate about the potential benefits and risks that may result from the use of such crops.

It is essential that we improve food production and distribution in order to feed and free from hunger a growing world population, while reducing environmental impacts and providing productive employment in low-income areas. This will require a proper and responsible utilisation of scientific discoveries and new technologies. The developers and overseers of GM technology applied to plants and micro-organisms should make sure that their efforts address such needs.

Foods can be produced through the use of GM technology that are more nutritious, stable in storage and in principle, health promoting—bringing benefits to consumers in both industrialised and developing nations. New public sector efforts are required for creating transgenic crops that benefit poor farmers in developing nations and improve their access to food through employment-intensive production of staples such as maize, rice, wheat, cassava, yams, sorghum, plantains and sweet potatoes. Cooperative efforts between the private and public sectors are needed to develop new transgenic crops that benefit consumers, especially in the developing world.

Concerted, organised efforts must be undertaken to investigate the potential environmental effects, both positive and negative, of GM technologies in their specific applications. These must be assessed against the background of effects from conventional agricultural technologies that are currently in use. Public health regulatory systems need to be put in place in every country to identify and monitor any potential adverse human health effects of transgenic plants, as for any other new variety.

Private corporations and research institutions should make arrangements to share GM technology, now held under strict patents and licensing agreements, with responsible scientists for use for hunger alleviation and to enhance food security in developing countries. In addition, special exemptions should be given to the world's poor farmers to protect them from inappropriate restrictions in propagating their crops.

Today there are some 800 million people (18% of the population in the developing world) who do not have access to sufficient food to meet their needs, primarily because of poverty and unemployment. Malnutrition plays a significant role in half of the nearly 12 million deaths each year of children under five in developing countries. In addition to lack of food, deficiencies in micro-nutrients (especially vitamin A, iodine and iron) are widespread. Furthermore, changes in the patterns of global climate and alterations in use of land will exacerbate the problems of regional production and demands for food. Dramatic advances are required in food production, distribution and access if we are going to address these needs. Some of these advances will occur from non-GM technologies, but others will come from the advantages offered by GM technologies.

Achieving the minimum necessary growth in total production of global staple crops—maize, rice, wheat, cassava, yams, sorghum, potatoes and sweet potatoes—without further increasing land under cultivation, will require substantial increases in yields per acre. Increases in production are also needed for other crops, such as legumes, millet, cotton, rape, bananas and plantains. It is important to increase yield on land that is already intensively cultivated. However, increasing production is only one part of the equation. Income generation, particularly in low-income areas together with the more effective distribution of food stocks, are equally, if not more, important. GM technologies are relevant to both these elements of food security.

In developing countries, it is estimated that about 650 million of the poorest people live in rural areas where the local production of food is the main economic activity. Without successful agriculture, these people will have neither employment nor the resources they need for a better life. Farming the land, and in particular smallholder farming, is the engine of progress in the rural communities, particularly of less developed countries.

The domestication of plants for agricultural use was a long-term process with profound evolutionary consequences for many species.

One of its most valuable results was the creation of a diversity of plants serving human needs. Using this stock of genetic variability through selection and breeding, the 'Green Revolution' produced many varieties that are used throughout the world. A good example of a selective breeding was the introduction of 'dwarf' genes into rice and wheat which, in conjunction with fertiliser applications, dramatically increased the yield of traditional food crops in the Indian sub-continent, China and elsewhere. Despite past successes, the rate of increase of food crop production has decreased recently (yield increase in the 1970s of 3% per annum has declined in the 1990s to approximately 1% per annum).

There are still heavy losses of crops owing to biotic (e.g. pests and disease) and abiotic (e.g. salinity and drought) stresses. The genetic diversity of some crop plants has also decreased and there are species without wild relatives with which to cross-breed. There are fewer options available than previously to address current problems through traditional breeding techniques though it is recognised that these techniques will continue to be important in the future. Increasing the amount of land available to cultivate crops, without having a serious impact on the environment and natural resources, is a limited option.

Modern agriculture has increased production of food, but it has also introduced large-scale use of pesticides and fertilisers that are expensive and can potentially affect human health or damage the ecosystem. A major challenge faced by humankind today is how to increase world food production and people's access to food, which requires local and employment-intensive staples production, without further depleting non-renewable resources and causing environmental damage. Research on transgenic crops, as with conventional plant breeding and selection by farmers add or remove a character of choice in a plant, bearing in mind the regional needs and opportunities.

It offers the possibility off not only bringing in desirable characteristics from other varieties of the plant, but also of adding characteristics from other unrelated species. Thereafter the transgenic plant becomes a parent for use traditional breeding. Modification of qualitative and quantitative characteristics such as the composition of protein, starch, fats or vitamins by modification of metabolic pathways has already been achieved in some species. Such modifications increase the nutritional status of the foods and may, help to improve human health by addressing malnutrition and under-nutrition. GM technology has also shown its potential to address micronutrient deficiencies and thus reduce the national expenditure and resources required to implement the current supplementation programmes.

These nutritional improvements have rarely been achieved previously by traditional methods of plant breeding. Transgenic plants with important traits such as pest and herbicide resistance are most necessary where no inherent resistance has been demonstrated within the local species. There is intense research on the development of resistance to viral, bacterial, and fungal diseases; modification of plant architecture (eg height) and development (eg early or late flowering or seed production); tolerance to abiotic stresses (eg salinity and drought); production of industrial chemicals (plant-based renewable resources); and the use of transgenic plant biomass for novel and sustainable sources of fuel.

Other benefits from transgenic plants under study include increased flexibility in crop management, decreased dependency on chemical insecticides and soil disturbance, enhanced yields, easier harvesting and higher proportions of the crop available for trading. For the consumer this should lead to decreased cost of food and higher nutritive value.

A large proportion of developing world agriculture is in the hands of small-scale farmers whose interests must be taken into account. Concerns regarding GM technology range from its potential impact on human health and the environment to concerns about private sector monopolies of the technology. It is essential that such concerns are addressed if we are to reap the potential benefits of this new technology.

Examples of GM Technology

GM technology has been used to produce a variety of crop plants to date, primarily with 'market-led' traits, some of which have become commercially successful. Developments resulting in commercially produced varieties in countries such as the USA and Canada have centred on increasing shelf-life of fruits and vegetables, conferring resistance to insect pests or viruses, and producing tolerance to specific herbicides. While these traits have had benefits for farmers, it has been difficult for the consumers to see any benefit other than, in limited cases, a decreased price owing to reduced cost and increased ease of production.

A possible exception is the development of GM technology that delays ripening of fruit and vegetables, thus allowing an increased length of storage. Farmers would benefit from this development by increased flexibility in production and harvest. Consumers would benefit by the availability of fruits and vegetables such as transgenic tomatoes modified to soften much more slowly than traditional varieties, resulting in improved shelf-life and decreased cost of production, higher quality and lower cost.

It is possible that farmers in developing countries could benefit considerably from crops with delayed ripening or softening as this may allow them much greater flexibility in distribution than they have at present. In many cases small-scale farmers suffer heavy losses due to excessive or uncontrolled ripening or softening of fruit or vegetables. The real potential of GM technology to help address some of the most serious concerns of world agriculture has only recently begun to be explored. The following examples show how GM technology can be applied to some of the specific problems of agriculture indicating the potential for benefits.

Pest Resistant Crops

There is clearly a benefit to farmers if transgenic plants are developed that are resistant to a specific pest. For example, papaya-ringspot-virus-resistant Papaya has been commercialised and grown in Hawaii since 1996. There may also be a benefit to the environment if the use of pesticides is reduced. Transgenic crops containing insect resistance genes from Bacillus thuringiensis have made it possible to reduce significantly the amount of insecticide applied on cotton in the USA.

For example, one analysis showed a reduction of five million acre-treatments (two-million-hectare-treatments) or about one million kilograms of chemicals insecticides in 1999 compared with 1998. However, populations of pests and disease-causing organisms adapt readily and become resistant to pesticides, and there is no reason to suppose that this will not occur equally rapidly with transgenic plants.

In addition, pest biotypes are different in various regions. For instance, insect resistant crops developed for use in the USA and Canada may be resistant to pests that are of no concern in developing countries, and this is true both for transgenic plants and those developed by conventional breeding techniques. Even where the same genes for insect or herbicide resistance are useful in different regions, typically these genes will need to be introduced into locally adapted cultivars. There is need, therefore, for more research on transgenic plants that have been made resistant to local pests to assess their sustainability in the face of increased selection pressures for ever more virulent pests.

High Yield Crops

One of the major technologies that led to the 'Green Revolution' was the development of high-yielding semidwarf wheat varieties. The genes responsible for height reduction were the Japanese NORIN 10 genes introduced into Western whcats in the 1950s. These genes had two benefits: they produced a shorter, stronger plant that could respond to more fertiliser without collapsing, and they increased yield directly by reducing cell elongation in the vegetative plant parts, thereby allowing the plant to invest more in the reproductive plant parts that are eaten.

These genes have recently been isolated and demonstrated to act in exactly the same way when used to transform other crop plant species. This dwarfing technique can now potentially be used to increase productivity in any crop plant where the economic yield is in the reproductive rather than the vegetative parts.

Biotic and Abiotic Stresses

The development of crops that have an inbuilt resistance to biotic and abiotic stress would help to stabilise annual production. For example, Rice Yellow Mottle Virus (RYMV) devastates rice in Africa by destroying the majority of the crop directly, with a secondary effect on any surviving plants that makes them more susceptible to fungal infections. As a result this virus has seriously threatened rice production in Africa. Conventional approaches to the control of RYMV using traditional breeding methods have failed to introduce resistance from wild species to cultivated rice.

Researchers have used a novel technique that mimics 'genetic immunisation' by creating transgenic rice plants that are resistant to RYMV. Resistant transgenic varieties are currently entering field trials to test the effectiveness of their resistance to RYMV. This could provide a solution to the threat of total crop failure in the sub-Saharan African rice growing regions.

Numerous other examples could be given to illustrate the range of current scientific research including transgenic plants modified to combat papaya ring spot virus, blight resistant potatoes and rice bacterial leaf blight; or as an example of an abiotic stress, plants modified to overproduce citric acid in roots and provide better tolerance to aluminum in acid soils . These examples have clear commercial potential but it will be imperative to maintain publicly funded research in GM technology if their full benefits are to be realised. For example, while GM technology provides access to new gene pools for sources of resistance, it needs to be established that these sources of resistance will be more stable than the traditional intra-species sources.

Marginalised Land

A vast landmass across the globe, both coastal as well as terrestrial has been marginalised because of excessive salinity and alkalinity. A salt tolerance gene from mangroves has been identified, cloned and transferred to other plants. The transgenic plants were found to be tolerant to higher concentrations of salt. The gutD gene from Escherichia coli has also been used to generate salt-tolerant transgenic maize plants. Such genes are a potential source for developing cropping systems for marginalised lands.

Nutritional Benefits

Vitamin A deficiency causes half a million children to become partially or totally blind each year. Traditional breeding methods have been unsuccessful in producing crops containing a high vitamin A concentration and most national authorities rely on expensive and complicated supplementation programmes to address the problem. Researchers have introduced three new genes into rice: two from daffodils and one from a micro-organism. The transgenic rice exhibits an increased production of beta-carotene as a precursor to vitamin A and the seed is yellow in colour. Such yellow, or golden, rice may be a useful tool to help treat the problem of vitamin A deficiency in young children living in the tropics.

Iron fortification is required because cereal grains are deficient in essential micro-nutrients such as iron. Iron deficiency causes anaemia in pregnant women and young children. About 400 million women of childbearing age suffer as a result and they are more prone to stillborn or underweight children and to mortality at childbirth. Anaemia has been identified as a contributing factor in over 20% of maternal deaths (after giving birth) in Asia and Africa.

Transgenic rice with elevated levels of iron has been produced using genes involved in the production of an iron-binding protein and in the production of an enzyme that facilitates iron availability in the human diet. These plants contain 2 to 4 times the levels of iron normally found in non-transgenic rice, but the bioavailability of this iron will need to be ascertained by further study.

Environmental Impact Reduction

Water availability and efficient usage have become global issues. Soils subjected to extensive tillage (ploughing) for controlling weeds and preparing seed beds are prone to erosion, and there is a serious loss of water content. Low tillage systems have been used for many years in traditional communities. There is a need to develop crops that thrive under such conditions, including the introduction of resistance to root diseases currently controlled by tillage and to herbicides that can be used as a substitute for tillage.

Applications in more developed countries show that GM technology offers a useful tool for the introduction of root disease resistance for conditions of reduced tillage. However, a careful cost-benefit analysis would be needed to ensure that maximum advantage is achieved. Regional differences in agricultural systems and the potential impact of substituting a traditional crop with a new transgenic one would also need to be carefully evaluated.

First generation transgenic varieties have benefited many farmers in the form of reduced production costs, higher yields, or both. In many cases, they have also benefited the environment because of reduced pesticide usage or by providing the means to grow crops with less tillage. Insects are responsible for huge losses to crops in the field and to harvested products in transit or storage, but health concerns for consumers and for environmental impact have limited the registration of many promising chemical pesticides.

Genes for pest resistance carefully deployed in crops to avoid selecting for future pest resistance, provides alternative opportunities to reduce the use of chemical pesticides in many important crops. In addition, lowering the contamination of our food supply by pathogens that cause food safety problems (eg mycotoxins) would be beneficial to farmers and consumers alike.

Pharmaceuticals and Vaccines from Transgenic Plants

Vaccines are available for many of the diseases that cause widespread death or human discomfort in developing countries, but they are often expensive both to produce and use. The majority must be stored under conditions of refrigeration and administered by trained specialists, all of which adds to the expense. Even the cost of needles to administer vaccines is prohibitive in some countries.

As a result, the vaccines often do not reach those in most need. Researchers are currently investigating the potential for GM technology to produce vaccines and pharmaceuticals in plants. This could allow easier access, cheaper production, and an alternative way to generate income. Vaccines against infectious diseases of the gastro-intestinal tract have been produced in plants such as potato and bananas. Another appropriate target would be cereal grains.

An anti-cancer antibody has recently been expressed in rice and wheat seeds that recognises cells of lung, breast and colon cancer and hence could be useful in both diagnosis and therapy in the future. Such technologies are at a very early stage in development and obvious concerns about human health and environmental safety during production must be investigated before such plants can be approved as speciality crops. Nevertheless, the development of transgenic plants to produce therapeutic agents has immense potential to help in solving problems of disease in developing countries.

About one third of medicines used today are derived from plants, one of the most famous examples being aspirin (the acetylated form of a natural plant product, salicylic acid). It is believed that less that 10% of medicinal plants have been identified and characterised, and the potential exists to use GM technology in a way that increases yields of these medicinal substances once identified.

For example, the valuable anti-cancer agents vinblastine and vincristine are the only approved drugs for treatment of Hodgkin's lymphoma. Both products are derived from the Madagascar Periwinkle, which produces them in minute concentrations along with 80 to 100 very similar chemicals. The therapeutic compounds are therefore extremely expensive to produce. Currently, there is intensive research in progress to investigate the potential of GM technology to increase the yields of active compounds, or to allow their production in other plants that are easier to manage than the Periwinkle.

Transgenic crop research and development should focus on plants that will (i) improve production stability; (ii) give nutritional benefits to the consumer; (iii) reduce the environmental impacts of intensive and extensive agriculture; and (iv) increase the availability of pharmaceuticals and vaccines; while (v) developing protocols and regulations that ensure that transgenic crops designed for purposes other than food, such as pharmaceuticals, industrial chemicals, etc. do not spread or mix with either transgenic or non-transgenic food crops.

Transgenic Plants in Human Health

Through classical plant breeding techniques, present day cultivated crops have become significantly different from their wild counterparts. Many of these crops were originally less productive and at times unsuitable for human consumption. Over the years, traditional plant breeding and selection of these crops have resulted in plants that are more productive and nutritious. To date, over 30 million hectares of transgenic crops have been grown and no human health problems associated specifically with the ingestion of transgenic crops or their products have been identified.

However numerous potential concerns have been raised since the development of GM technology in the early 1970s. Such concerns have focused on the potential for allergic reactions to food products, the possible introduction or increase in production of toxic compounds as a result of the GM technology, and the use of antibiotic resistance as markers in the transformation process. Every effort should be made to avoid the introduction of known allergens into food crops.

Information concerning potential allergens and natural plant toxins should be made available to researchers, industry, regulators, and the general public. In order to facilitate this effort, public databases should be developed which facilitate access of all interested parties to data. Traditional plant breeding methods include wide crosses with closely related wild species, and may involve a long process of crossing back to the commercial parent to remove undesirable genes.

A feature of GM technology is that it involves the introduction of one or at most, a few, well-defined genes rather than the introduction of whole genomes or parts of chromosomes as in traditional plant breeding. This makes toxicity testing for transgenic plants more straightforward than for conventionally produced plants with new traits, because it is much clearer what the new features are in the modified plant. On the other hand, GM technology can introduce genes from diverse organisms, some of which have little history in the food supply.

Decisions regarding safety should be based on the nature of the product, rather than on the method by which it was modified. It is important to bear in mind that many of the crop plants we use contain natural toxins and allergens. The potential for human toxicity or allergenicity should be kept under scrutiny for any novel proteins produced in plants with the potential to become part of food or feed. Health hazards from food, and how to reduce them, are an issue in all countries, quite apart from any concerns about GM technology.

Since the advent of GM technology, researchers have used antibiotic resistance genes as selective markers for the process of genetic modification. The concern has been raised that the widespread use of such genes in plants could increase the antibiotic resistance of human pathogens. Kanamycin, one of the most commonly used resistance markers for plant transformation is still used for the treatment of the following human infections: bone, respiratory tract, skin, soft-tissue, and abdominal infections, complicated urinary tract infections, endocarditis, septicaemia, and enterococcal infections.

Scientists now have the means to remove these marker genes before a crop plant is developed for commercial use. Developers should continue to move rapidly to remove all such markers from transgenic plants and to utilise alternative markers for the selection of new varieties. No definitive evidence exists that these antibiotic resistance genes cause harm to humans, but because of public concerns, all those involved in the development of transgenic plants should move quickly to eliminate these markers.

Transgenic Plants and the Environment

Modern agriculture is intrinsically destructive of the environment. It is particularly destructive of biological diversity, notably when practised in a very resource-inefficient way, or when it applies technologies that are not adapted to environmental features (soils, slopes, climatic regions) of a particular area. This is true of both small-scale and large-scale agriculture. The widespread application of conventional agricultural technologies such as herbicides, pesticides, fertilisers and tillage has resulted in severe environmental damage in many parts of the world. Thus the environmental risks of new GM technologies need to be considered in the light of the risks of continuing to use conventional technologies and other commonly used farming techniques.

Some agricultural practices in parts of the developing world maintain biological diversity. This is achieved by simultaneously cultivating several varieties of a crop and mixing them with other secondary crops, thus maintaining a highly diverse community of plants. Most of the environmental concerns about GM technology in plants have derived from the possibility of gene flow to close relatives of the transgenic plant, the possible undesirable effects of the exotic genes or traits (eg insect resistance or herbicide tolerance), and the possible effect on non-target organisms.

As with the development of any new technology, a careful approach is warranted before development of a commercial product. It must be shown that the potential impact of a transgenic plant has been carefully analysed and that if it is not neutral or innocuous, it is preferable to the impact of the conventional agricultural technologies that it is designed to replace. Given the limited use of transgenic plants world-wide and the relatively constrained geographic and ecological conditions of their release, concrete information about their actual effects on the environment and on biological diversity is still very sparse.

As a consequence there is no consensus as to the seriousness, or even the existence, of any potential environmental harm from GM technology. There is therefore a need for a thorough risk assessment of likely consequences at an early stage in the development of all transgenic plant varieties, as well as for a monitoring system to evaluate these risks in subsequent field tests and releases.

Risk assessments need base-line information including the biology of the species, its ecology and the identification of related species, the new traits resulting from GM technology, and relevant ecological data about the site(s) in which the transgenic plant is intended to be released. This information can be very difficult to obtain in highly diverse environments. Centres of origin or diversity of cultivated plants should receive careful consideration because there will be many wild relatives to which the new traits could be transferred.

For special environments, transgenic plants can be developed using technologies that minimise the possibilities of gene flow via pollen and its effects on wild relatives, through the use of male sterility methods or maternal inheritance resulting from chloroplast transformation. Studies of gene transfer from conventional and transgenic plants to wild relatives and other plants in the ecosystem have so far concentrated on species of economic importance such as wheat, oilseed rape and barley. A virtual absence of data, particularly for species like maize, imposes the need to carefully and continuously monitor any possible effects of novel transgenic plants in the field. In addition there is a continued need for research on the rates of gene transfer from traditional crops to indigenous species.

The risk assessments performed should be standardised for plants new to an environment. Most nations already have procedures for the approval and local release of new varieties of crop plants. Although these assessments are based primarily on the agronomic performance of the new variety compared with existing varieties, this approval process could serve as the beginning or model for a more formal risk assessment process to investigate the potential environmental impact of the new varieties, including those with transgenes.

Historically, both poverty and structural change in rural areas have resulted in severe environmental deterioration. The adoption of modern biotechnology should not accelerate this deterioration. It should instead be used in a way that reduces poverty and its deleterious effects on the environment.

The public sector and charitable foundations funded the national and international crop research in the post-war period that led to doubling or tripling crop yields in large parts of Asia and Latin America, along with gains in employment and nutrition in the developing world. The dwarf wheat and rice plants and other high-yielding varieties which were at the centre of this 'Green Revolution' met the needs of millions of poor farmers and consumers.

The balance of funding of this kind of research has shifted significantly during the past decade from the public to the private sector, and there has been a corresponding reduction in national, non-commercial agricultural research capacity that needs to be reversed. Substantial public-sector agricultural research still exists, however, notably in North America, Australia, Europe, China, India, Brazil, and in the Consultative Group for International Agricultural Research (CGIAR) system.

The CGIAR system comprises 16 international research centres with interests that include wheat and maize (Mexico), rice (Philippines), potatoes (Peru), and millet and sorghum (India), but the financial support for the CGIAR has been declining in real terms. Whereas fundamental research is still being carried out in the public sector, the strategic application, in sharp contrast to the 'Green Revolution', takes place largely in the private sector where much of the intellectual property is controlled.

In these circumstances, research priorities are driven by market forces (eg price signals). Companies produce products whose costs are recoverable in the market place. There are also goods that benefit society as a whole rather than individuals and whose costs cannot be recovered in the market place (so-called public goods). Public sector funding is needed for such public good work. A classic example of a public good would be an improved plant that can be propagated by farmers with little deterioration, as with self-pollinated (eg wheat and rice), or vegetatively propagated (eg potatoes) crops. If such crop improvement research were left to normal markets for private provision, then it would be systematically under-supplied.

The main reason why aid donors and foundations support international agricultural research is to ensure that public-good research of relevance to small-scale farmers and to complex tropical and subtropical environments is undertaken. If such research were wholly private, even in a perfectly functioning market, the demands of rich consumers for innovation in their own interests would overwhelm both the needs of and the price signals from poor consumers and small-scale farmers.

Given the limited resources so far available to them for research, the non-commercial (public and charitable foundation) sectors have achieved more than could have been expected (eg beta-carotene enhanced rice and rice resistant to RYMV).

Capacity Building

The development of a strong capability in the plant sciences is an absolute priority for all national research programmes. This is necessary because only local plant breeding can address local agricultural environments and only local initiatives can appreciate cultural preferences. It is very likely that genes and gene functions will, as our understanding grows, be increasingly transferable between crops and between agricultural environments. However, in order for these genes to be incorporated into adapted, tested, safe and effective varieties, a sustained local research capacity will be necessary. This is equally true whether the genes are transferred by GM technology or by conventional breeding.

The International Research Centres (eg, those under CGIAR sponsorship) and national research programmes must use partnerships with advanced research institutions to increase their efforts to extend the new GM technologies to crops such as bananas, plantains, beans, sorghum, wheat, maize, cassava and potatoes, which are important resources for many countries. These centres should also take a lead in developing alliances with advanced and strategic research institutions, both public and private, in order to ensure the transfer of the appropriate technologies. In addition, the new communications technologies should be vigorously supported to facilitate the free exchange of knowledge and best practices among the world's agricultural research and farming communities.

If world agriculture, and developing countries in particular, are to benefit from the many potential advantages of GM technology, it will be important to promote capacity building in risk management. In order to be effective, the following objectives must be included:

— Build sufficient scientific and technical human resources in each country to enable it to assess the relative benefits and the risks of GM technology;

— Strengthen local and global infrastructure;

— Monitor and evaluate the short mid and long-term effects of transgenic plants and share data between all relevant countries;

— Develop simple techniques to readily and reliably distinguish non-transgenic and transgenic plants where necessary.

Intellectual Property

Current industrial biotechnology is primarily orientated to the needs of large-scale commercial agriculture, rather than to those of the subsistence farmer. Most developing countries lack the financial resources and are limited in the scientific infrastructure needed to develop their own biotechnology programmes for the crops that are important to feed their people. The long-term decline of public agricultural research, the increasing privatisation of GM technologies, and the growing emphasis on the crops and priorities of the industrialised nations do not bode well for feeding the increasing populations of the developing world.

Without changed incentives for sharing access to GM technologies, the world is unlikely to direct much of its research for improved nutrition and employment-based access to staples for the poor. The application of modern genomics research techniques to plant species promises an explosion of new knowledge and information that may lead to important new advances in agricultural production and the quality, quantity and variety of food products. Actual realisation of these advances will depend to a significant extent on both publicly and privately funded research and on the development efforts of commercial companies supported by private investment.

As in other areas of biotechnology, intellectual property rights are likely to play an important role in securing economic returns for the intellectual and financial investments that make the research and developments possible. An important consideration regarding such intellectual property rights in inventions and discoveries resulting from genomic research and from other applications of biotechnology is that overly broad intellectual property rights should not be granted. To grant such rights would stifle further research and development. Intellectual property rights should be narrowly tailored to be commensurate with the actual scope of new inventions and discoveries so as not to impede continuing research, innovation and development.

To benefit the growing populations of the developing world, new plant varieties will have to be developed through a variety of sources, including:

i) farmers who select plants that succeed best in their particular locality for the retention of seed for future use or sale;

ii) public or pro bono research institutions financed out of taxes or charitable grants that provide improved varieties to appropriate users free or at cost; and

iii) for-profit companies interested in creating new products and markets that develop new varieties financed through profits from seed sales.

As instruments of public policy, intellectual property regimes should facilitate the maximum possible innovation in development of beneficial new crop varieties through individual, public and corporate sources, as well as promote research collaboration.

Special attention should be paid to international conventions that may affect innovation in agriculture. These conventions include Trade Related Intellectual Property (TRIPs), patent law, plant variety protection and the Convention on Biological Diversity (CBD). To be effective, these conventions should be consistent with each other so as to reduce any distortions in the promotion of innovation by farmers, public research institutions, and for-profit corporations.

At present, it appears that many less developed countries are reluctant to join in international intellectual property agreements on plants because they believe that such agreements will create a system that strongly favours the corporate sector—while simultaneously hampering the public and private sector efforts that support their own citizens. In fact, many of the intellectual property rights that exist today in industrialised nations apply to the tools used in research and development to produce new transgenic plant varieties. If the rights to these tools are strongly and universally enforced—and not extensively licensed or provided pro bono in the developing world—then the potential applications of GM technologies described previously are unlikely to benefit the less developed nations of the world for a long time.

Private companies today can obtain plant varieties free of charge from farmers and from non-commercial institutions such as the CGIAR, add one or more proprietary traits, and then release seed with a variety of forms of legal or technical protection against copying, farm retention, or farm-to-farm transmission. Thus, a market-based system exists, based in part on non-reimbursed contributions from farmers and institutions such as the CGIAR. This heavily concentrates advances in research within companies whose legitimate search for profit naturally fails to focus their research on poverty and long term sustainability issues. Transgenic plants have intensified the dilemma because a high level of skill and infrastructure is needed to develop them.

Moreover, broad patents have been granted to companies that secure their competitiveness in the market place. To remain viable, public-sector research by farmers, the CGIAR, and by National Agricultural Research Systems needs to be strengthened and provided with increased resources and attention—both from governments and from the world's agricultural scientists. In addition, intellectual property rights should be obtained by these public-sector institutions for their discoveries so that these rights can be used in negotiations with the private sector to produce increased public benefit.

Intensive agriculture requires the use of certified seed (ie seed free of pathogens, pests, and weeds) and growers purchase new seed every year as an established practice. Most growers plant hybrid varieties of maize (corn) and other crops that are more uniform and vigorous than ordinary varieties because of heterosis (hybrid vigour) and these advantages are lost when second generation seed is used. In addition, some growers are under contract with food processors who demand specific quality standards that require new seed to be purchased annually.

However, for some crops (eg soybeans) many growers save seed to plant in subsequent years (seed re-use), until reduced yields induce them to buy new seed. Saving seed is often not an optimal practice for reasons related to the contamination of seed with pests and pathogens. In developing countries, government programmes often attempt to provide clean seed at affordable prices. However, in many instances, small growers cannot afford to purchase new seed every year, and they wish to maintain their long-standing practice of saving some of the seed from one year's crop in order to plant next year's crop.

Historically, fertility and reproduction of grain crops in Africa, Asia and parts of the Americas have acquired a deep spiritual significance. Seeds are exchanged freely and are given away to travellers from far-away lands. It is clear that growers in developing counties feel strongly that it is their right to decide whether to use their own seed or purchase fresh, certified seed.

The general public would seem to be very much on the side of the growers on this issue. To ensure financial return for their investments, many biotechnology seed companies have sought to prevent the use of second-generation seed produced from transgenic crops. For example, growers who purchase transgenic plant seeds are often required to sign contracts that specifically prohibit the saving and replanting of second-generation seed.

Over the long-term, the most significant form of intellectual property protection for seeds may prove to be technological. A specific example of this that has been the source of much controversy is a patent application for an invention whereby traits in transgenic plants would be expressed only if a certain chemical activator was applied to seeds or plants (Genetic Use Restriction Technology, GURT). This technology involves the use of a chemical treatment of seeds or plants that either inhibits or activates specific genes involved in germination. One technology involves a complex three-gene system whereby one gene produces a protein that interferes with proper plant embryo development, thus preventing seed germination.

The expression of this gene is allowed by applying tetracycline (or other chemicals) which prevents a recombinase gene from being repressed by an induced protein. Once the recombinase is expressed after tetracycline application, a blocking sequence placed between a transiently-active promoter and the inhibitor gene is removed, thus allowing the expression of the protein that prevents germination. The seed sold to farmers would be pre-treated with either tetracycline or other chemicals (copper, steroids, etc).

Most experts agree that there are considerable technical problems yet to be solved and that GURT will not be available for commercialisation for several years. The possible commercialisation of GURT technology for controlling the use of transgenic plant seeds has generated considerable public debate, being referred to as 'terminator technology'. On the one hand, growers, especially in developing countries, maintain their right to retain and plant second-generation seeds. On the other, the seed companies seek to obtain a return on investment so that they can continue to invest in new technologies. Both parties, as well as the general public, have an important stake in these issues. There is a clear need for a resolution that serves the wider public interest.

In an alternative GURT, the transgenic traits would be expressed only if a certain chemical activator were applied to seeds or plants. In this case, farmers would retain the ability to save their own seed, yet lack access to the added traits in the absence of payment for chemical activators. GURTs potentially have beneficial applications for consumers, growers, and the environment that should not be overlooked in debates over intellectual property rights. For example, GURTs could be used to prevent transgenes from spreading to closely-related wild plants by preventing germination of any crossbred seeds. Furthermore, this technology could potentially eliminate the problems of 'volunteer' plants that appear from seed left in the field after harvest.

Volunteer plants must be eliminated before the next crop is planted because they are hosts for pests and pathogens and can nullify the benefits of crop rotation. As with any growth regulator applied to crops, there are possible environmental or human health issues associated with the use of chemical activators (ie tetracycline, copper, steroids) and these would need to be addressed. Other concerns regarding the use of GURT are economic, related to the intellectual property rights and the monopoly of production of transgenic plants by particular companies. It is critical that the potential benefits of GM technology become available to developing countries.

2

POTENTIALS OF CROP BIOTECHNOLOGY

New molecular or recombinant DNA methodologies, also called biotechnology or genetic engineering, allow the breeder to modify plants in different ways. The creation of genetically engineered plants depends on a unique characteristic of plants that makes it is possible to take cells from any part of a plant and, with appropriate "coaxing," make them undergo sustained division and ultimately give rise to an entire plant. If you can introduce a new gene into a single plant cell and identify which one it is, that single cell can give rise to an entire plant, each cell of which contains the information, the new DNA.

Technically how is this done? A part of a plant is removed, e.g., leaf, stem, seed, and then DNA is introduced into a small subset of the cells in that plant tissue. The introduction can be done either biologically or physically. The biological method involves the use of a bacterium known as Agrobacterium, which naturally infects plant cells and can, as a natural consequence of its infection, insert DNA into the plant's genetic material. Scientists learned how to put genes of interest into the inserted DNA, thereby having a method to stably introduce a new gene.

Another method of DNA introduction involves microscopic, DNA-coated "bullets" propelled at high speeds, which ultimately end up inside the cell, where the DNA comes off the "bullets" and becomes a heritable part of the plant. Once DNA is introduced, the geneticist propagates the cells. During that growth period it is necessary to identify cells that have stably maintained the new information. This can be done through the introduction of genetic information that confers a selective advantage on the engineered cell, e.g. an herbicide- or antibiotic resistance gene or a gene that allows cells to grow under nutritional conditions that the non-engineered cell cannot.

The next challenge is to "coax" the cells to reform a plant, through the manipulation of hormones and other components of the growth medium. Depending on the plant species, this can be the most difficult part of the processes. However, once accomplished, you have a plant, each cell of which contains the new information.

What can be done with these technologies? Some applications utilise the molecular tools but don't involve changing the genetic makeup of the plant. In one application, the tools are used to speed up the process of breeding by providing molecular "road" maps that tell the breeder what parts of the genome have been kept in the offspring. Other molecular tools are used to detect diseases in farmer's fields at an early stage to help control pest infestation and reduce pesticide use.

Other uses of the technology involve the actual introduction of information into the plant, some from the plant itself and some from other sources. Some of these products have already entered the marketplace, e.g., the NewLeaf potato, BollGard cotton, Roundup Ready soybean, and YieldGard corn. Other products are only in the very early stages of development, many in university laboratories. These include plants with the following characteristics:

— Increased pest resistance to fungal, bacterial, viral and insect pests,

— Improved agronomic characteristics, such as increased yield and a heightened ability to utilise fertilisers and withstand stresses,

— Better post-harvest/processing characteristics, like slow- or no- ripening fruits and vegetables, and higher solids tomatoes,

— Reduction of antinutrients, such as removal of glycoalkaloids or allergens,

— Improved nutritional characteristics, e.g. enrichment for antioxidants and micronutrients, folic acid, iron,

— Pharmaceutical benefit, e.g. plants that vaccinate or prevent/ treat human and animal disease,

— Provision of alternative sources of industrial oils, fuels, and plastics, and

— Remediation capability for metal and organic pollutants from soils and water.

As with other technologies developed in the past, e.g. the domestication of plants and animals, agricultural mechanisation, chemical fertilisers, pesticides, these new genetic tools bring questions about risks and benefits. While few if any activities in today's technologically complex world involve zero risk, individuals are looking to minimise human and environmental risk. We must be educated about these technologies and their uses and be informed participants in the dialogue about how best to utilise them.

Biotechnology in Agriculture

Cross Breeding, Hybridasation and Bioengineering

Scientists have been improving plants by changing their genetic makeup since the late 1800s. Typically, this has hybridisation accomplished through cross-breeding and hybridisation, in which two related plants are cross-fertilised and the resulting offspring have characteristics of both parent plants. In the breeding process, however, many undesirable traits often can appear in addition to the desirable ones. Some of those undesirable traits can be eliminated through additional breeding, which is time consuming.

Breeders can then further select and reproduce the offspring that have the desired traits.

Many of the foods that are already common in our diet are obtained from plant varieties that were developed using conventional genetic techniques of breeding and selection. Hybrid maize, nectarines, which are genetically altered peaches, and tangelos, which are a genetic hybrid of a tangerine and grapefruit, are all examples of such breeding and selection. Today, by inserting one or more genes into a plant, scientists are able to produce a plant with new, advantageous characteristics.

The new gene splicing techniques are being used to achieve many of the same goals and improvements that plant breeders historically have sought through conventional methods. They give scientists the ability to isolate genes and introduce new traits into foods without simultaneously introducing undesirable traits. This is an important improvement over traditional breeding. Because of the increased precision offered by the bioengineered methods, the risk of introducing detrimental traits is actually likely to be reduced.

Food Safety

The Food and Drug Administration (FDA) has found no evidence to indicate that either ordinary plant deoxyribonucleic acid (DNA) or the DNA inserted into plants using bioengineering presents food safety problems. Nor are the small amounts of the newly expressed proteins likely to change dramatically the safety profile of the plant. If safety concerns should arise, however, they would most likely fall into one of three broad categories: allergens, toxins or anti-nutrients.

FDA has extensive experience in evaluating the safety of such substances in food. It is important to note that the kinds of food safety testing typically conducted by developers of a bioengineered food crop to ensure their foods meet all applicable requirements of the Food, Drug and Cosmetics Act (FD&C Act) address these potential concerns. In the event that something unexpected does occur, this testing provides a way to detect such changes at the developmental stage and defer marketing until any concern is resolved.

As aforementioned, some of the food safety concerns that could arise include:

— *Allergens*: Foods normally contain many thousands of different proteins. While the majority of proteins do not cause allergic reactions, virtually all known human allergens are proteins. Since genetic engineering can introduce a new protein into a food plant, it is possible that this technique could introduce a previously unknown allergen into the food supply or could introduce a known allergen into a "new" food.

— *Toxins*: It is possible that a new protein, as introduced into a crop as a result of the genetic modification, could cause toxicity.

— *Anti-nutrients*: It is possible that the introduction of antinutrients, such as molecules like phytic acid, could reduce essential dietary minerals such as phosphorus.

The use of genetic engineering techniques could also result in unintended alterations in the amounts of substances normally found in a food, such as a reduction of Vitamin C or an increase in the concentration of a naturally occurring toxicant in the plant food.

Bioengineered foods and food ingredients must adhere to the same standards of safety under the FD&C Act that apply to their conventionally bred counterparts. This means that these products must be as safe as the traditional foods in the market. FDA has the power to remove a food from the market or sanction those marketing the food if the food poses a risk to public health. It is important to note that the FD&C Act places a legal duty on developers to ensure that the foods they market to consumers are safe and comply with all legal requirements.

Food Additive

A substance that is intentionally added to food is a food additive, unless the substance is generally recognised as safe (GRAS) or is otherwise exempt, such as a pesticide whose safety is overseen by Environmental Protection Agency (EPA). The FD&C Act requires premarket approval of any food additive regardless of the technique used to add it to food. Thus, substances introduced into food are either new food additives that require premarket approval by FDA, or GRAS and are therefore exempt from the requirement for premarket review.

Generally, foods such as fruits, vegetables and grains are not subject to premarket approval because they have been safely consumed over many years. Other than the food additive system, there are no premarket approval requirements for foods generally. Under FDA policy, a substance that would be a food additive if it were added during traditional food manufacturing is also treated as a food additive if it is introduced into food through bioengineering of a food crop.

Our authority permits us to require premarket approval of any food additive and, thus, to require premarket approval of any substance intentionally introduced via bioengineering that is not generally recognised as safe. Examples of substances intentionally introduced into food that would be reviewed as food additives include those that have unusual chemical functions, have unknown toxicity, or would be new major dietary components of the food. For example, a novel sweetener bioengineered into food would likely require premarket approval. In our experience with bioengineered food to date, however, we have reviewed only one substance under the food additive provisions, an enzyme produced by an antibiotic resistance gene, and we granted it approval as a food additive.

In general, substances intentionally added to or modified in food via biotechnology to date have been proteins and fats that are, with respect to safety, similar to other proteins and fats that are commonly and safely consumed in the diet and, thus, are presumptively GRAS. Therefore, they have not needed to go through the food additive approval process.

Pre-Mark Consultation

Typically, the consultation begins early in the product development stage, before the product is ready for market. Company scientists and other officials meet with FDA scientists to describe the product they are developing. The agency then advises the company on what tests would be appropriate for the company to assess the safety of the new food. FDA evaluates the information for all of the known hazards and also for potential unintended effects on plant composition and nutritional properties since plants may undergo changes other than those intended by the breeders. For example, FDA scientists are looking to assure that the newly expressed compounds are safe for food consumption and that there are no allergens new to the food, no increased levels of natural toxicants, and no reduction of important nutrients.

They are also looking to see whether the food has been changed in any substantive way such that the food would need to be specially labelled to reveal the nature of the change to consumers. If a plant developer used a gene from a source whose food is commonly allergenic, FDA would presume that the modified food might be allergenic. The developer, however, is allowed the opportunity to demonstrate that such food would not cause allergic reactions in persons allergic to food from the source.

Labelling: One of the most important issues confronting the biotechnology industry is that of labelling. Under the FD&C Act, a food is misbranded if its labelling is false or misleading in any particular way. FDA does not require labelling to indicate whether or not a food or food ingredient is a bioengineered product, just as it does not require labelling to indicate which conventional breeding technique was used in developing a food plant. However, if genetic modifications materially change the composition of a food product, these changes must be reflected in the food's labelling. This would include its nutritional content (for example, more oleic acid or greater amino acid or lysine content) or requirements for storage, preparation or cooking, which might impact the food's safety characteristics or nutritional qualities. For example, one soybean variety was modified to alter the levels of oleic acid in the beans. Because the oil from this soybean is significantly different from conventional soybean oil, we advised the company to adopt a new name for that oil, a name that reflects the intended change.

If a bioengineered food were to contain an allergen not previously found in that food and if FDA determined that labelling would be sufficient to enable the food to be safely marketed, FDA would require that the food be labelled to indicate the presence of the allergen. FDA has received comments suggesting that foods developed through modern biotechnology should bear a label informing consumers that the food was produced using bioengineering.

In August 2002, USA President Bush's Office of Science and Technology Policy (OSTP) proposed strengthening controls over field trials to address the potential of material from field trials inadvertently getting into food or feed. FDA's task is to publish draft guidance for comment on procedures to address the possible intermittent, low-level presence in food and feed of new non-pesticidal proteins from biotechnology-derived crops that are under development for food or feed use but have not gone through FDA's premarket consultation process.

Under this guidance, FDA would encourage sponsors. domestic and foreign, to submit protein safety information when field testing showed that there could be concerns that new non pesticidal proteins produced in the field-tested plants might be found in food or feed. FDA's focus would be on proteins new to such plants because FDA believes that at the low levels expected from such material, any food or feed safety concerns would be limited to the potential that a new protein could cause an allergic reaction in some people or could be a toxin.

Pharmaceutical Crops

FDA has the authority and responsibility for regulating pharmaceuticals, whether they are manufactured in a traditional manufacturing plant or manufactured in crops in the field. For crops in the field, however, there are additional issues to be addressed, including issues involving the parts of the plant that do not contain the pharmaceutical and the residual crop left over after a pharmaceutical is extracted.

In September 2002, FDA and USDA published Draft Guidance for Industry on the use of bioengineered plants or plant materials to produce biological products, including medical devices, new animal drugs, and veterinary biologics. This draft guidance outlines the important scientific questions and information that should be addressed to FDA by those who are using bioengineered plants to produce medical or veterinary products.

Agricultural Biotechnology in World Food Aid

Food aid is one of several global mechanisms created to deal with hunger and food insecurity. The need for food aid around the globe varies from specific responses to acute and episodic shortages to long-term donations of food to abate continuing chronic inability of some regions to become agriculturally self-sufficient. While agricultural biotechnology is not a panacea to food insecurity, it is likely to play a vital role in the delivery of food assistance and reduction of hunger for generations to come.

The U.N. Universal Declaration of Human Rights declares the right of access to food and freedom from hunger as a fundamental right. Although we live in a world of unprecedented prosperity and technological development, 800-850 million people are malnourished. More than 200 million of these are children, many of whom will never reach their full intellectual and physical potential. Another 1-1.5 billion humans have only marginally better access to food and often do not consume balanced diets containing sufficient quantities of all required nutrients.

The majority of this nutritionally at-risk population lives in developing countries. Most, perhaps 75 percent, live in rural agricultural regions. Most are very poor. There is a well-recognised link between poverty and hunger. In fact, family income is probably the single most important determinant of adequacy of access to food. The World Food Summit in 2002 reaffirmed a commitment made by the international community five years earlier to halve the number of hungry people by the year 2015.

That goal will not be met unless agricultural productivity and personal income can be improved in the world's poorest regions. It is argued by some that eliminating poverty is more important than producing more food since there is more than enough food produced in the world to feed everyone. Economists tell us that there is a surplus of food in the world—or at least a surplus of grain that when tabulated as potential caloric intake could theoretically adequately feed the current global population. But the sad lesson of both recent and ancient history is that adequate food supplies do not reach everyone.

The large number of hungry people proves that. It is pointless to argue whether poor agricultural productivity or extreme poverty is more to blame when people are starving. What is clear is that if the rural poor can produce a surplus of food in a more efficient and sustainable manner, there will be adequate food supplies, increasing income and the opportunity for supporting rural development. While most experts would agree that the only long-term solution to hunger is economic development and the elimination of poverty, people who are food self-sufficient through local or regional agriculture will not go hungry.

Unfortunately, neither the required increases in agricultural productivity nor the necessary rural development will happen overnight. The question then becomes. The shortterm solution for the hungry is food aid. But even food aid has become politicised as skeptics have charged that it is simply a way for rich over-producing nations to eliminate the surpluses produced by their heavily subsidised farmers. The skeptics also assert that food aid robs local farmers of markets and makes them hungrier. These arguments ignore the daily reality faced by hundreds of millions of hungry people for whom the immediate alternatives are simple: continued hunger and ultimate starvation or the acceptance of food aid.

Eliminating Chronic Hunger

The Green Revolution of the 1960s and 1970s helped India and China and other Asian countries become agriculturally self-sufficient net exporters of food in the last three decades. The increased productivity has been accompanied by increases in personal income and stimulus to national economies. Similarly, through application of new technology, agricultural productivity per hectare has doubled in most developed countries in the same timeframe.

The development of new high-productivity agricultural technologies resulted from investment in agricultural research performed in government laboratories, research universities, and non-governmental institutes such as the Consultative Group on International Agricultural Research (CGIAR) centers scattered around the globe. A crucial element of success has been the deployment of effective systems of out-reach education and technology transfer. Research and technology transfer has also taken place in the private sector.

For a variety of complex reasons, improvements in agricultural productivity did not take place in all developing countries. Quite the contrary, some of the least developed countries are now even less able to produce sufficient food. There, the Green Revolution never happened. While civil unrest and political corruption may have contributed greatly to this phenomenon, from an agricultural point of view, the failure lies in the lack of investment in and adoption of new technologies and management practices. Often this occurred because there was not sufficient attention paid or investment made in research to develop effective local or region-specific strategies and technologies.

Sub-Saharan Africa is a region where growth in agricultural production has not kept pace with expanding need. As a whole, the region has some of the poorest and most depleted agricultural soils. Only 4 percent of the farmed land is irrigated. Significant areas of agricultural land are at risk of becoming desert while in some parts of the region excessive humidity and high temperatures contribute to a high incidence of disease and pests. Weeds such as Striga stifle yields. Droughts are commonplace in some parts of the region. Outright crop failure is common and poor yields are endemic. There is clearly a need to develop crop varieties and management strategies that are more productive under these conditions.

High on the list of desired traits are crops with enhanced resistance to environmental stresses such as drought, temperature and salinity; enhanced resistance to diseases and pests; and improved agronomic properties and yield potential. The heavy reliance on a few staple crops makes biofortification—the boosting of the vitamin and mineral components of foods to enhance the nutritional value—an attractive strategy as well. Recent advances in molecular biology and genomics greatly enhance the plant breeder's capacity to introduce new traits into plants.

Commercial applications of agricultural biotechnology have already produced crops such as Btmaize, rice, potatoes, cotton and sweet corn (sweet maize) that can protect themselves against insects; virus-resistant papaya, squash and potatoes; and herbicide-tolerant crops such as wheat, maize, sugar cane, rice, onions and beets that allow more effective weed management. There is accumulating evidence that these biotech crops can be more productive and profitable for farmers. Major reductions in costs for labour, energy and chemicals have been documented.

The crops have also proven to be environmentally-friendly, particularly with regard to biodiversity, reduction of agricultural chemicals in soil and water, and decreased exposure of workers and communities to chemicals. There is also an emerging international consensus of scientific and regulatory opinion that crops derived through biotechnology are safe to eat as food and feed and beneficial for the environment. These and other promising technologies are now being directed at improving the production and yield of African staple crops: banana, cassava, maize, millets, oil crops, peanut, potato, rice, sorghum, soybean, sweet potato and wheat. Protein-enhanced sweet potatoes and potatoes and carotene-enhanced rice and oilseeds promise to improve the nutritional value of the diet.

Thus, over the long term, agricultural biotechnology promises to play a crucial role in improving agricultural productivity and reducing the environmental impact of agriculture leading to agricultural sustainability and food security in many regions of the world. While it would be foolish to say that agricultural biotechnology alone will solve the world's food problems, it would be equally foolish to assert that food insecurity can be eliminated without agricultural biotechnology.

In recent years, there has been a significant change in the organisation of agricultural research directed at improving food security. It is now recognised that research needs to be done at local, national and regional levels in order to address specific agricultural challenges and produce new varieties appropriate to local agriculture and customs. This change is particularly focused on utilising and expanding local scientific and agricultural human and capital infrastructure that can work in partnership with international scientists and funding. Although the path is clear and there are numerous successful examples of these kinds of international partnership, global funding for such activities falls far short of the level required to achieve global food security in the next decades.

Widespread local or regional crop failure often leads to acute food shortages and hunger. The reason for episodic events can be as varied as flood, droughts or civil war. The United Nations, national governments and an assortment of nongovernmental organisations (NGOs) often respond by mobilising an immediate food aid programme. Food aid distribution can be hindered by lack of infrastructure for storage and transportation of food, and there are often concerns for the security of aid workers.

Recently, a new obstacle to food aid distribution has been identified. Repeated crops failures in Southern Africa have placed millions of people in six nations at risk. In response, the United States offered food aid that included substantial shipments of maize. The maize supply in the United States is approximately 30-35 percent insect-protected Bt-maize developed through biotechnology. This variety of maize had been approved by the U.S. Environmental Protection Agency (EPA), the U.S. Department of Agriculture (USDA) and the Food and Drug Administration (FDA) as safe for consumption as food and feed. It was commingled with conventional maize in the U.S. commodity system.

However, since the intended recipient nations did not use biotech seed varieties and imported few commodities such as maize, they for the most part lacked specific laws and regulatory systems with respect to foods produced through biotechnology. Genetically modified (GM) maize was an unapproved food in their regulatory systems. In light of the global scare campaign against GM foods, several countries hesitated to accept the aid. Ultimately, intensive international consultation and fact-finding satisfied all of these countries save Zambia, which continued to refuse GM food aid. One obvious conclusion to be drawn from this experience is that regulatory systems and training need to be in place before the need for food aid arises again.

Development of Plant Biotechnology

Corn (maize) originated in Mexico from a grass called teosinte that has a small reproductive structure bearing little resemblance to the ear of corn seen in markets around the world today.

Tomatoes and potatoes first appeared in South America—tomatoes as small fruits the size of a grape, and potatoes as knobby tubers with high concentrations of a family of bitter chemicals called glycoalkaloids, which are toxic to humans. Through selective breeding by our ancestors, the shape, colour and chemical content of these and hundreds of other plants consumed today have been modified to suit consumer preference or to obtain desired characteristics such as high yield, disease and insect resistance, and tolerance to drought and other plant stresses.

Not only have these plants changed in appearance and composition, they also have become distributed worldwide through centuries of human migration and trade. For example, cabbage, which originated in Europe, is now grown on every inhabited continent. When today's consumers walk into a market in many parts of the world, they are witnesses to today's global food system where foods produced in one part of the world are daily shipped to local markets.

Ancient ancestors were modifying the genetic makeup of plants by transferring genetic material from one plant to another. However, it wasn't until Gregor Mendel, an Austrian monk, conducted experiments in the 1800s with garden peas that the basic laws of heredity were first unravelled. Prior to the early 1900s, traditional plant breeding, like that practiced by Mendel, relied on man-made artificial crosses in which pollen from one plant species was transferred to another sexually compatible plant. The goal was to take a desirable trait from one plant and introduce it into another plant.

However, often desirable characteristics either were not present in sexually compatible plants or did not occur in any plant species. This led plant breeders to seek new ways of transferring desirable genes. Beginning in the 1930s, plant breeders developed techniques to allow them to develop plants from two parent plants that could not normally produce viable offspring. An example is the technique called "embryo rescue," in which the new plant embryo is provided with extra care in the laboratory to enable it to survive during its early growth. In the 1950s, plant breeders also developed methods of creating variation in an organism's genetic structure through what is termed "mutation breeding."

Mutations in the genetic makeup of a plant occur continuously and randomly in nature through such events as the sun's radiation and may lead to the occurrence of new desirable traits. Mutation breeding uses similar random processes to cause changes in a plant's genes. Plants then are assessed to determine if the genes were changed and whether the changes provided a beneficial trait such as disease or insect resistance. If the plant was "improved," then it was tested for other changes that may have occurred.

Many of the common food crops we use daily have been developed through techniques such as embryo rescue and mutation breeding and virtually all the foods we consume have genes in them. It is hard to think of an example of a common food crop in the developed world that has not been improved by some form of modern technology, or what can be termed "biotechnology." Simply put, biotechnology is a set of techniques that utilises living organisms, or parts of organisms, to make or modify products, improve plants or animals, or develop microorganisms for specific purposes. This definition encompasses all human activities conducted on living organisms from the earliest development of plant breeding 10,000 years ago to the present. This is the reason plant breeders consider the term "genetically modified organisms"—or GMOs—to be misnomer since all common food crops of today have been so modified.

Modern Genetic Engineering: Genetic engineering is one form of biotechnology and usually refers to copying a gene from one living organism plant, animal or microbe—and adding it to another organism. In genetic engineering, a small piece of genetic material (DNA) is inserted into another organism to produce a desired effect. This is in contrast to traditional plant breeding in which all the genes desirable and undesirable contained in the male plant—pollen—are combined with all the genes of the female plant. The progeny resulting from this cross may contain the gene for the desirable character, but it will also contain many of the undesirable genes from both parents.

Genetic engineering has the advantage of being able to transfer only the gene of interest and greatly accelerate plant breeding. But genetic engineering also is more powerful than traditional breeding since it can move genes not only between similar plant species but also from distant relatives, including non-plant species. It is possible to move genes between such seemingly unrelated organisms because all living organisms share the same code for DNA and the synthesis of proteins and other basic life functions. What might seem on the surface to be very different organisms are, in fact, very similar, at least at the molecular level.

All living things are more alike than different, and this is one of the reasons that genes can be moved so successfully between such seemingly different organisms as plants and bacteria. Genes are not unique to the organisms from which they came, so there really aren't "tomato genes" or "bacterial genes." It's the collection of all genes in a tomato or a bacterium that makes it a tomato or bacterium, not a single gene. Most plant species differ by only a small percentage of their genes and that even such seemingly different organisms as tomatoes and bacteria have many common genes. These findings suggest that in the long-term evolutionary process even tomatoes and bacteria had some common ancestor.

From the discovery 50 years ago of the structure of DNA, scientists soon came to realise they could take segments of DNA that carried information for specific traits—genes—and move them into another organism. In 1972, the collaboration of Hubert Boyer and Stanley Cohen resulted in the first isolation and transfer of a gene from one organism to a single-celled bacterium where it would express the gene and manufacture a protein. Their discoveries led to the first direct use of biotechnology—the production of synthetic insulin to treat people with diabetes—and the start of what is often called modern biotechnology.

Plants were first transformed through genetic engineering in the late 1970s. Mary-Dell Chilton and colleagues used a common soil-dwelling bacterium, Agrobacterium tumefaciens, that attaches itself to plants and transfers some of its DNA into the plant. Chilton and her colleagues added a gene to this bacterium, which in turn transferred the gene into a plant where it became part of the plant's DNA. This bacterium is still commonly used in genetic engineering along with another technique that uses a high-velocity mechanism to inject DNA into plant cells. The result from either technique is the same—the plant cells take up the gene and begin to express it as their own.

Plants developed through genetic engineering were first grown on 1.7 million hectares in 1996 in the United States, but by 2002 they were grown on 58.7 million hectares in 16 countries. By far the major use of the present plants is to manage pests—weeds, insects and diseases. Weed management with genetically engineered plants is accomplished because the plants have a modified enzyme (a protein) that allows them to survive an application of a specific herbicide that normally acts on that enzyme.

Growers can plant the herbicide-tolerant seeds, allow the plants to emerge along with any weeds in the field and then treat the field with an herbicide. The result is that the weeds, but not the crops, die. The advantage to growers is that they spend less time on weed management, have enhanced weed control, use safer herbicides, and in many cases use less herbicides. Additionally, this technology allows growers to use soil conservation practices such as reduced or no-tillage, thus helping to retain soil structure and moisture and reduce erosion. Herbicide tolerant crops (soybean, canola, cotton and maize) were grown on 48.6 million hectares in 2002.

Insect-resistant crops developed through genetic engineering utilise the common soil bacterium, Bacillus thuringiensis (Bt), which has been commercially used for more than 50 years, as an insecticide spray. Although safe to humans and the environment, when a susceptible insect ingests Bt, the Bt protein binds to specific molecular receptors in the gut and creates pores causing the insect to starve to death. Insecticidal products containing Bt were first commercialised in France in the late 1930s, but even in 1999 the total sales of Bt products constituted less than 2 percent of the total value of all insecticides. Bt, which had limited use as a foliar insecticide, became a major insecticide only when genes that produce Bt toxins were engineered into major crops.

The Bt crops available at present are maize and cotton. These were grown on a total of 14.5 million hectares in 2002. Virus-resistant crops were created by inserting a non-infective part of a plant virus into a plant, essentially "vaccinating" the plant to protect it from the virus. This technique is called "pathogen-derived resistance." Squash and papaya have been engineered to resist infection by some common viruses and are approved for sale in the United States. There are fewer than 1 million hectares of these crops. The bioengineered plants available at present provide growers with better tools to manage pest problems.

As with any technology, there are risks and benefits to currently available genetically engineered plants, but the present body of information indicates their use has enhanced pest management, substantially reduced the amounts of pesticides used in some crops, enabled growers to use safer pesticides, and contributed to enhanced safety for humans and the environment. The regulatory process for managing these plants and their effects on the environment and humans has evolved with the technology and the scientific community's knowledge of these tools.

Many of the more controversial issues surrounding genetic engineering of plants—such as pesticide resistance, gene flow and intellectual property issues—are not unique to this new technology but pertain to all types of agriculture. Some species of insects have developed resistance to sprays of Bt, indicating the potential for some species to become resistant to Bt plants. However, despite Bt plants being grown on more than 62 million hectares worldwide from 1996 to 2002, there have been no documented cases of resistance development.

The reasons for this lack of resistance appear to involve not only biological factors of the insects and the Bt plant, but also the fact that the regulatory agency (the Environmental Protection Agency) in the United States requires a resistance management plan for growing Bt plants. No other insecticide has such strict regulations. Still, growers, companies and federal regulatory agencies must be vigilant about resistance developing for biotech crops used to manage insects, weeds and viruses as they also must with non-biotech pest management tactics. It will be important to consider the accrued environmental and health benefits of these biotech crops prior to the development of any resistance and how resistance can be managed if and where it occurs.

In addition to pesticide resistance, gene flow from biotech to non-biotech crops may also be a concern. However, the risk of gene flow varies with each crop and each gene. Pollen flow in soybeans is very limited so the risk of a biotech soybean crop crossing with a non-biotech soybean is minimal, but this may be different for another crop. Likewise, if the gene in the biotech crop that provided a pest management trait, such as insect resistance, moved into a non-biotech plant, such as a weed, any selective advantage of the insect-protected weed in the ecosystem should be assessed. These same questions should also be answered with non-biotech crops, but these have not received the same level of attention as biotech crops because of the latter's higher profile.

In the future, the potential uses of plant biotechnology are far more wide-ranging than the pest-management biotech crops of today. Plants are being developed that serve as production "factories" for medically important drugs, sources of alternative energy, tools for cleaning toxic waste sites, and biomaterials including dyes, inks, detergents, adhesives, lubricants, plastics and the like. Consumers may see these products as more directly enhancing their quality of life than the pest-management biotech crops of today.

Perhaps an even more dramatic advantage to consumers will be seen when plants are genetically engineered to have enhanced health benefits such as disease-fighting chemicals or increased amounts of essential vitamins and minerals. A healthy and well-informed discussion of the risks and benefits involved in agricultural biotechnology is needed to ensure a proper role for this new technology in our future food and health systems. No one should believe that any technology, including biotechnology, will completely solve the world's agricultural problems. Many people familiar with biotechnology, however, believe it to be an important component of the solution.

Global Communication Ecology and Biotechnology

Today similarly charged stories are told about genetically modified (GM) foods. In addition to claims about the negative impact of GM foods on the environment and human health, there are wild claims that associate GM foods with maladies such as brain cancer and impotence as well as behavioural changes. Some of these rumours are spread at the highest levels of government in developing countries. The tactics employed in the debates are equally sophisticated. Critics of the technology have used instruments of mass communication to provide the public with information that is carefully designed to highlight the dangers they attribute to biotechnology.

Advocates of biotechnology have often been forced to respond to charges against the technology and have only on rare occasions taken the initiative to reach out to the public. This is particularly important because the general public does not readily understand the technical details of biotechnology products and so new communication approaches are needed. While advocates of biotechnology have often tried to rely on the need for scientific accuracy, critics employ rhetorical methods that are designed to invoke public fear and cast doubt on the motives of the industry. The critics draw analogies between the "dangers" of biotechnology with the catastrophic consequences of nuclear power or chemical pollution.

Indeed, they use terms like "genetic pollution" and "Frankenstein foods". Critics have also relied on the general distrust of large corporations among sections of the global community to make their case. In addition, they have made effective use of incidents, whose risks they have amplified. In a study, researchers indicated that pollen from GM maize producing a Bt toxin killed the larvae of Monarch butterflies. This study was used to dramatise the impact of biotechnology on the environment. Subsequent published peer explanations of the limitations of the study and refutations of the conclusions did not change the original impression created by the critics of biotechnology.

In this case the real environmental issue was not if GM maize killed monarch butterfly larvae or not. The critical question was what impact the maize had on the environment compared to maize grown with chemical pesticides. It is the issue of relative risks that is important; not simply a single event examined outside the wider ecological context. But apparently, this kind of analysis would not serve the cause of critics. It is notable that the critics of biotechnology have defined the rules of the debate in two fundamental ways.

First, they have managed to create the impression that the onus of demonstrating safety lies with advocates of biotechnology and not on its critics. In other words, biotechnology products are considered unsafe until proven otherwise. Second, they have been effective in framing the debate in environmental, human health and ethical terms, thereby masking the underlying international trade considerations. By doing so, they have managed to rally a much wider constituency of activists who are genuinely concerned about environmental protection, consumer safety and ethical social values.

There is a general view that concerted efforts to promote public debate will improve communication and lead to the acceptance of biotechnology products. This may be the case in some situations. But generally, the concerns are largely material and cannot be resolved through public debate alone. This is mainly because the root causes of the debate lie in the socio-economic implications of the technology and not mere rhetorical considerations. It is possible that public debates will only help to clarify or amplify points of divergence and do little to address fundamental economic and trade issues.

Operating in the new global communication ecology will require greater diversity of biotechnology products, an increase in the number of institutional players, enhanced policy research on life sciences and society, and stronger policy leadership.

Much of the debate on the role of biotechnology in developing countries is based on hypothetical claims with no real products in the hands of producers or consumers. Under such circumstances, communication and dialogue are not enough until there is a practical reference point. In other words, rebutting the claim of critics is not as important as presenting the benefits of real products in the market place.

This can best be achieved through collaborative efforts among local scientists, entrepreneurs, policy makers and legitimate civil society organisations. There is ample evidence to suggest that concerns over the safety of new products tend to decline as local participation and ownership in new technologies increase. Similarly, local participation in new technologies increases the level of trust in new technologies, thereby reducing the demand for non-science-based safety regulations. For example, the word of a farmer from South Africa stating the positive impact of GM cotton on her welfare carries more weight than thousands of screaming press releases and empty headlines on both sides of the debate.

This means that spreading the use of biotechnology not only promotes familiarity with the technology, but also generates the information needed to convince the public about the relevance and usefulness of the technology. The broadening of the range of products is therefore a key aspect of the debate. This is particularly important in developing countries interested in using the technology to enhance local products and diversify their food base.

Information on the development of drought-tolerant crops, for example, would be relevant to African countries while other regions might be interested in different products. This view also suggests that general debates about the role of biotechnology are of little utility unless framed in the context of local needs and applications. The absence of a real stake in the technology creates a vacuum that is often filled with misinformation on the risks and benefits of the technology. Countries such as Kenya and South Africa that have their own biotechnology research programmes have a more considered view of the technology.

Addressing the issue of biotechnology communication requires an improved understanding of the changing ecology of communication. The ecology includes a complex network of sources of information and opinion leaders as well as new communication tools that were hitherto not available to the public or advocacy groups. In his days, Khair Beg was outraged to learn that coffeehouses had become an authoritative source of information on what was happening in his jurisdiction.

Similarly, the Internet has become a more important communication tool than classical methods such as TV advertising. But unlike in the days of Khair Beg, the new communication ecology is global in character, making it possible to spread information widely and generate empathy among a diversity of activist organisations, including those that are unlikely to be affected by the technology. These cyber-communities are built around a complex set of mailing lists that are not easily accessible. Correcting misinformation spread through such channels is difficult to do because of the complexity of the networks. While activists tend to use a diverse array of social movements to advance their cause, advocates have tended to focus on the use of centralised institutions whose impact is largely negligible in the modern communication ecology.

But creating the necessary diversity requires a broadening of the base of social movements that champion the role of science and technology in human welfare. One of the most important aspects of the biotechnology debate has been the role of the popular media. In Europe, for example, the media have played an important role in amplifying claims by critics or creating doubt about positions advanced by advocates of the technology.

In contrast, support for the role of science does not usually have the polemical turn that newspaper editors relish. The traditional view that science is based on immutable facts which can be passed on from an authority to the general public is being challenged by approaches that demand greater participation in decision-making. In other words, scientific information is being subjected to democratic practices. The debate over biotechnology has pushed the frontiers of public discourse of technical matters. On the one hand, society is being forced to address issues that are inherently technical, and on the other, the scientific community is under pressure to accept non-technical matters as valid inputs to decision-making.

3

Transformation and Regeneration of Plants

The discovery of DNA transformation in bacteria and the realisation that genetic information is encoded in the nucleotide base sequence of DNA suggested that it might be possible to transform higher organisms with DNA-coding sequences to effect directed genetic changes. One of the first successful methods harnessed a circular DNA molecule, or plasmid, carried by the pathogenic plant bacterium Agrobacterium tumefaciens. The bacterium invades wounds in plant tissue, where it stimulates rapid host cell growth, resulting in the production of a plant tumour or gall.

This occurs because a tumour-inducing (Ti) segment of the plasmid DNA becomes integrated in a chromosome of the host cell nucleus. Deleting the genes that control the production of tumour cells from the plasmid leaves right and left border fragments, each 24 nucleotides long, that mark the ends of the integrating Ti segment. This segment of bacterial plasmid DNA can be used as a vector to introduce foreign DNA. If a bacteriophage lambda cos site is included in the vector, up to ≈40 kb of DNA can be packaged in vitro between the borders. Transformation vectors include marker genes that can be detected in bacterial and plant host cells and controlling elements, or promoters, to drive the expression of the desired gene in plant tissue at the appropriate phase of development.

A drawback of the Ti vector system is that it is restricted to plants that are hosts for Ag. tumefaciens. Another method uses protoplasts that are prepared by treating plant cells with enzymes that dissolve cell walls. Held in a suitable osmoticum, such as a 4-mollL solution of mannitol, they are stable and do not burst. The protoplasts are treated with a solution of transforming DNA, some of which is taken up and may become integrated in the host cell genome. The uptake of DNA can be increased by treatment with polyethylene glycol or by electrical disruption of the protoplast membrane. The protoplasts must form new cell walls as the osmoticum is gradually replaced with culture medium.

Embryogenic cell lines are selected from the resulting unorganised callus that is formed. The biolistic, or gene gun, method uses very small DNA coated metal particles (gold or tungsten) that are rapidly accelerated toward a target of plant cells or tissues. The particles penetrate the plant cell walls and some enter the cell nuclei. In a small proportion of these cells the introduced DNA becomes integrated in a host chromosome. A selectable marker is needed to recover putative transformants.

Plant Transformation

In the context of food-crop improvement several benefits may be achieved by strain modification. High yields reduce the costs of raw materials, quality improvements can reduce processing costs, and agronomic characteristics such as herbicide tolerance and resistance to pests and diseases may reduce both farm costs and agrochemical residues in raw materials. Novel characteristics that add value to raw materials can be introduced by transformation.

Herbicide Tolerance

Research on the mechanisms of herbicide toxicity led to the isolation of genes that encode target enzymes that are not inhibited by the agent. Incorporating these specific genes into crops makes it possible to use nonselective herbicides with desirable qualities such as a short half-life in the environment and very low or no toxicity to nontarget organisms, especially mammals, on a wide range of crops. The herbicide Roundup inhibits the enzyme 5-enolpyruvyl-shikimate-3-phosphate synthase, which is involved in the synthesis of aromatic amino acids in the chloroplast.

Until a gene for glyphosate tolerance was available, Roundup was used to keep land free from plant growth or to prepare soil for crop cultivation. It is rapidly decomposed in the soil and thus has little residual toxicity. Genes for herbicide tolerance provide an alternative to chemical screening tests for identifying selective herbicides. The sulfonyl urea and imidasolinone herbicides interfere with amino acid biosynthesis in plants by inhibiting the enzyme acetohydroxyacid synthase (AHAS), which is responsible for the synthesis of leucine and valine. Mutations at two different sites in the AHAS gene create enzyme forms that are resistant to both types of herbicides.

Phosphinothricin (PPT) inhibits glutamine synthetase in plants. The bar gene, from Streptomyces hygroscopicus, encodes an enzyme, phosphinothricin acetyltransferase, that converts PPT to a nontoxic derivative, thus conferring resistance to PPT herbicides in transgenic plants. An alternative method of creating herbicide-tolerant crop forms is to select herbicide-tolerant mutants from plant cell tissue cultures. Herbicide-tolerant corn lines recovered from tissue culture mutants were used to make F_1 hybrids tolerant to the herbicide Pursuit.

Pest Resistance

The development of transformation as a breeding tool has increased interest in identifying genes for resistance to pests and diseases. Attempts to clone the resistance genes already in use by breeders proved to be unexpectedly difficult and breakthroughs have occurred only within the past year. The soil bacterium Bacillus thuringiensis forms a delta endotoxin protein. Ingested by the larva of a sensitive insect, the toxin induces a loss of the semipermeability of the intestinal wall so that hemolymph leaks into the intestinal cavity. The larva loses fluids and dies.

Several endotoxin proteins have been identified that are active specifically against different groups of insects. The cryl protein, active against many lepidopteran larvae, was first used as the biological insecticide BT. Suspensions of B. thuringiensis resting cells containing the cryl protein were shown to be safe and effective in large-scale applications over many years.

Expressed in transgenic crops, the Bt gene that encodes the cryl protein is effective in controlling the larvae of hornworm, cabbage looper, diamondback moth, and leaf miner. Because the toxin gene came from a prokaryote, the codon usage had to be adjusted to maximise expression in a eukaryote. Transgenic plants that express Bt genes may have a short life in agriculture because of the strong selection pressure they exert for insects that are insensitive to the endotoxin protein. Resistance to BT was reported in diamondback moths after extensive use of this insecticide.

Special management practices such as using several different control methods together and reducing the level of expression of the Bt gene or the numbers of plants carrying it may be necessary if BT is to remain a useful biological control agent. The seeds of an exotic collection of cowpea germplasm that is unusually resistant to the larvae of moths that attack stored grain were found to contain a trypsin inhibitor. The gene that encodes this inhibitor was isolated and cloned from the cowpea accession and was introduced into transgenic plants. By interfering with the digestion of the plant proteins ingested by insect larvae feeding on the plants, transgenic plants that express this gene show moderately high resistance. Trypsin inhibitors and other protease inhibitors are commonly found in food and feed crops.

Virus Resistance

Many plant viruses are transmitted by insects that feed by

sucking plant sap. When the coat protein gene for the RNA virus that causes tobacco mosaic was introduced into the host plant its expression interfered with the replication and systemic spread of the virus, making the plant resistant. This phenomenon has also been observed in other plant viruses and their hosts. Naturally occurring virus resistance is uncommon, and the discovery showed much promise. The first commercial application of this discovery is the use of viral coat protein genes of cucumber mosaic, zucchini yellow, and watermelon mosaic viruses in squash.

The engineered squash contains viral coat proteins that account for <0.1% of the total protein of the squash fruit. This compares favourably with the somewhat greater virus content of commonly consumed fruit from plants that are naturally infected with virus. Other forms of virus resistance include viral genes that when expressed in plants interfere with replication of the viral genome or impede virus movement from cell to cell. The pokeweed plant produces several ribosornal inhibitor proteins. One of these expressed in transgemc plants protects them against a broad range of viruses.

Fungal Pathogens

Many fungal pathogens have cell walls made of chitin. Novel transgenic plants that express bean or Serratia marcescens chitinase genes have been reported to show resistance. Four genes for resistance to different plant pathogens were recenfly cloned from higher plants. Their structural similarity-all had regions of DNA with repeat lysine encoding motifs-suggests that they have a role in signaling biochemical defense responses. An earlier example was of a corn gene that appears to encode the structure of a membrane receptor protein to which a fungal pathotoxin binds. General methods for recovering and redeploying such genes will help to reduce our dependence on pesticides.

Applications of Recombinant DNA Technology in Food

Controlled Ripening

The first food product from this recombinant DNA technology, genetically engineered chymosin (rennin), is widely used in Europe and the United States for cheese making. The first genetically engineered food plant to be introduced in the United States, the Flavr Savr tomato (Calgene, Davis, CA), was released early in 1994. These tomatoes incorporate a Ti vector that carries an antisense form of the tomato gene that encodes polygalacturonase (PG).

Produced during fruit ripening, this enzyme breaks down pectin, the cement that holds the cells of the fruit tissue together, and is responsible for fruit softening. The PG antisense gene reduces the rnRNA transcript formed and hence the amount of PG. The result is a fruit with a longer shelf life at full ripeness. Because transformation is inefficient and very few treated cells are transformed, it is necessary to have a marker to select cells that carry integrated and expressed foreign DNA. The Ti vector thus carries a gene that encodes neomycin phosphotransferase H, which confers resistance to the antibiotic kanamycin. Kanamycin kills untransformed tomato cells.

The kanamycin resistance gene, driven by a constitutive promoter, is always expressed. The PG antisense gene is expressed only after flowering and fertilisation, late in plant development. The Food and Drug Administration concluded that the expression of the neomycin phosphotransferase II gene is of no significance for consumers of the fruit. An alternative means of delaying fruit ripening is by cosuppression. The addition of sense copies of a gene by transformation often results in the failure of the resident gene to be expressed. The mechanism of cosuppression is not fully understood. Cosuppression is also being used to delay ripening in tomatoes, a process that is regulated by the production of ethylene.

In this case, additional copies of the gene that encodes the enzyme 1-aminocyclopropane- 1-carboxylate synthose, which is important in ethylene synthesis, blocks ethylene production by cosuppression. The produce industry routinely triggers the ripening of fruit that are harvested before they are ripe by treating them with ethylene gas in storage. Antisense, cosuppression, or both can also be used to control the amounts of other materials that plants produce. The candidates include raffinose and stachyose, oligosaccharides in legume seeds (beans and soybean) that are responsible for flatulence. Naturally occurring variants of the bean (Phaseolus vulgaris) with low concentrations of stachyose have been used to produce bean cultivars that cause less flatulence.

Phytase Production

Genetic engineering promises improvements in the nutritional value of livestock feed. Phytate (myoinositol hexaphosphate) is the principal storage form of phosphorus in plant seeds. Inorganic phosphate is released from phytate by an endogenous enzyme, phytase, at seed germination. However, when plant seeds are fed to monogastric animals, such as pigs and poultry, these animals cannot use phytate and thus cannot access the phosphorus. After the oil has been extracted from oilseed rape (canola) in Europe the meal is used as animal feed with inorganic phosphate added as a supplement. Phytase, prepared from Aspergillus niger, can replace the phosphate supplement.

A gene from As. niger encodes the enzyme was used to transform tobacco with an agrobacterium vector. Phytase-expressing plants were normal in growth and development and their seeds were enriched in phytase. Seeds from transgemc tobacco were tested in feeding trials with male broiler chicks. This supplemented diet was cornpared with diets supplemented with and without inorganic phosphate and As. niger phytase. Growth in animals fed diets containing milled transgenic phytase-expressing seeds was significantly greater than that in animals fed control seeds or diets without supplementation. The production of transgenic oilseed rape and soybean that constitutively overexpress phytase in their seeds is now under way.

Phytase added to soil increases the availability of phosphorus to plant roots. These findings suggest that the expression of the enzyme in the roots of transgenic plants may have a similar effect. Reducing phytate in soybean and other vegetable products may play a role in human nutrition. Phytate is an effective chelator of divalent cations, such as iron and zinc. Consuming protein sources that contain relatively high concentrations of phytate results in decreased mineral bioavailability, and consuming vegetable proteins with low concentrations of phytate promotes mineral absorption.

Phytochelatins

Another class of plant compounds with potential importance in human nutrition is the phytochelatins, a class of inducible peptides that bind metals and may determine the availability from plant foods of trace metals such as zinc and copper that are important in human nutrition. However, phytochelatins are also active in the uptake of toxic heavy metals. Phytochelatins contain four to six repeated units of -γ-glutamyl cystine. Cadmium sensitivity in Arabidopsis thaliana is caused by mutations that reduce the synthesis of glutathione, the substrate for phytochelatin biosynthesis. Selection for greater cadmium sensitivity might provide a method for reducing unwanted phytochelatin in food plants, thereby reducing their heavy metal content.

Expression of Antigens

The ability of gut mucosal cells to respond to antigens has prompted attempts to create inexpensive oral vaccines by expressing antigens in transgenic food plants. Tobacco that was transformed with a gene encoding hepatitis B surface antigen expressed this protein in its leaves in a form that was antigenically similar to that in human serum. Because so much of our food is cooked and because heating may degrade the antigenic proteins, this approach will have limitations for human vaccination. There is also the prospect of engineering forage for livestock so that cow milk is induced to include antibodies that are effective against diseases of human infants, but, again, heating to sterilise or process such milk may render the antibodies ineffective.

Edible Plant Oils

Vegetable oils consist largely of triacylglycerols and their compositions vary considerably among genera and species. Plant breeding has exploited this variation to produce a variety of improved forms. The composition of canola oil was changed by breeding and selection to reduce the high amounts of erucic acid (22:1) that characterised early varieties. At the same time, glucosinolates (goitrogenic glycosides) were removed from canola seed so that the meal left after oil extraction in a crushing mill has greater value as livestock feed.

Increasing the value of plant edible oils in human nutrition by using cloned genes to alter the activities of specific enzymes responsible for lipid and fatty acid biosynthesis. The most common plant fatty acids have chain lengths of 16 or 18 carbon atoms and from 1 to 3 double bonds. They are liquid at room ternperature, and for making margarine, in which a higher melting temperature is required, the double bonds must be reduced by hydrogenation. However, this not only increases the saturated fat content of the oil but also converts many of the double bonds from the cis to the less desirable trans configuration.

Because the enzyme steraoyl-acyl carrier protein desaturase introduces the first double bond, it has been a target for antisense RNA. This approach was successful in two species of Brassica (B. napus and B. rapa) in which the content of stearic acid, which has no double bonds (18:0), in the seed oil was increased from 2% to 40%. Although vegetable oils have less saturated fatty acid than do animal fats, the 10-20% (depending on species) that they do contain is a target for further reduction by molecular biology. The major saturated fatty acid in plants is palmitic acid (16:0).

One approach depends on manipulating a branch point in fatty acid synthesis in which either the precursor palmitoyl-acyl carrier protein is elongated to form stearic acid or the acyl carrier protein is released to form palmitic acid. A reduction in the palmitic acid content was achieved by over expressing the enzyme that is responsible for elongation. Another successful approach was to use cosuppression to reduce the enzyme that releases palmitic acid.

Hypoallergenic Rice

Rice grains contain a 16-kDa globulin protein that is heat-stable and that resists proteolytic enzymes in the human gut. This protein is the cause of an ectopic dermatitis in sensitive Japanese children. The allergen can be destroyed by enzyme treatment, but the cost is excessive. Scientists recovered mutants in which the amount of allergenic protein was reduced after the rice seeds had been treated with a chemical mutagen. In one mutant with good agronomic qualities the allergenic protein content was reduced by ≈50%. Two other mutants with trace amounts of the allergen were largely sterile and were of little use in breeding and seed production. The allergen has been characterised and the gene that encodes it sequenced so it is theoretically possible to produce transgenic rice with an antisense gene or a cosuppressing construct that would be hypoallergenic.

Seed Proteins

Soybean and canola seed storage proteins, although important in human and livestock nutrition, could be of more value if their balance of amino acids were improved by increasing the methionine content. Used a gene from Brazil nut that encodes a

25 storage protein with a methionine content of 18% and introduced it by transformation into canola.

It increased the methionine content of the seed meal by >30%. The same gene expressed in the seeds of transgenic soybean also increased their nutritional quality but, unfortunately, made them allergenic to people who are sensitive to Brazil nuts. Evidently, the 2S protein is a major allergen in Brazil nuts. In view of this, it seems unlikely that the transgenic soybean will be commercialised in its present form.

Much attention is now paid in wheat breeding to the genes that encode high-molecular weight glutens and gliadins in the grain endosperm. Some of these proteins are responsible for the viscoelastic properties of doughs for making bread. Some progress has been made in cloning and expressing genes for low-molecular weight wheat storage proteins in tobacco seed. Engineering high-molecular weight proteins in wheat has yet to be achieved but once done offers prospects of improving the flour of other cereals, such as sorghum and millet, that are important in some developing countries.

Cruciferous Weed

Many of the spectacular advances made in molecular biology in the past 25 y have depended on the availability of a wide variety of model organisms, such as Escherichia coli, yeast, mice, Caenorhabditis elegans (a nonparasitic nematode), and corn. During this time there has been growing interest in the small cruciferous weed Ar. thaliana. This plant has a haploid number of four chromosomes and a DNA content of $\approx 0.1 \times 10^9$ nucleotide bases. Because this is much lower than the DNA content of crop plants, which is up to 100 times greater, there is a concerted effort to sequence the entire genome of this plant. Ar. thaliana has other advantages: it completes its life cycle from seed to seed in ≈6 wk, each flower produces several hundred seeds, and large plant populations can be grown in growth chambers or greenhouses in little space.

Genes of interest may be more readily identified and cloned from Ar. thaliana than from a crop plant. These genes may also be altered or their expression adjusted before they are used in an economically important plant. Ar. thaliana is thus a crop plant analogue that makes it possible for molecular biologists to try out systems that can later be applied to economic plants.

One of the first examples of such a use was the isolation of the acetolactate synthase gene from Ar. thaliana followed by the recovery of mutant forms that are insensitive to the herbicide imidasolinone. The resultant gene for tolerance could then be used to confer tolerance to unrelated crops after it had been introduced by transformation.

4

Genetic Engineering

Biotechnology is a broad term that applies to all practical uses of living organisms—anything from microorganisms used in the fermentation of beer to the most sophisticated application of gene therapy. The term covers applications that are old and new, familiar and strange, sophisticated and simple. The term is almost too broad to be useful.

One way of thinking about biotechnology is to consider two categories of activities: those that are traditional and familiar and those that are relatively new. Within each category can be found technologies that are genetic—that involve modifications of traits passed down from one generation to the next—and technologies that are not. Although there are interesting issues connected with a number of biotechnologies—both old and new—a new genetic biotechnology.

Genetic engineering refers to a set of technologies that are being used to change the genetic makeup of cells and move genes across species boundaries to produce novel organisms. The techniques involve highly sophisticated manipulations of genetic material and other biologically important chemicals. Genes are the chemical blueprints that determine an organism's traits. Moving genes from one organism to another transfers those traits. Through genetic engineering, organisms are given new combinations of genes—and therefore new combinations of traits—that do not occur in nature and, indeed, cannot be developed by natural means. Such an artificial technology is radically different from traditional plant and animal breeding.

Nature can produce organisms with new gene combinations through sexual reproduction. A brown cow bred to a yellow cow may produce a calf of a completely new colour. But reproductive mechanisms limit the number of new combinations. Cows must breed with other cows (or very near relatives). A breeder who wants a purple cow would be able to breed toward one only if the necessary purple genes were available somewhere in a cow or a near relative to cows. A genetic engineer has no such restriction. If purple genes are available anywhere in nature—in a sea urchin or an iris—those genes could be used in attempts to produce purple cows. This unprecedented ability to shuffle genes means that genetic engineers can concoct gene combinations that would never be found in nature.

Contrary to the arguments made by some proponents, genetic engineering is far from being a minor extension of existing breeding technologies. It is a radically new technology for altering the traits of living organisms by inserting genetic material that has been manipulated by artificial means.

Novel organisms bring novel risks, however, as well as the desired benefits. These risks must be carefully assessed to make sure that all effects—both desired and unintended—are benign. UCS advocates caution, examination of alternatives, and careful case-by-case evaluation of genetic enginering applications within an overall framework that seeks to move agricultural systems of food production toward sustainability.

Traditional Biotechnologies

A prime example of traditional genetic biotechnologies is selective breeding of plants and animals. The rudiments of selecting plants and animals with desirable traits and breeding them under controlled conditions probably go back to the dawn of civilisation, but the expansion of knowledge about genetics and biology in this century has developed selective breeding into a powerful and sophisticated technology. New molecular approaches like marker—assisted breeding (which enhances traditional breeding through knowledge of which cultivars or breeds carry which trait) promise to enhance these approaches even further.

Traditional breeding technologies have been immensely successful, and indeed are largely responsible for the high yields associated with contemporary agriculture. For multigene traits like intrinsic yield and drought resistance, they surpass genetic engineering. This is because selective breeding operates on whole organisms—complete sets of coordinated genes—while genetic engineering is restricted to three or four gene transfers with little control over where the new genes are inserted. For the most important agronomic traits, traditional breeding remains the technology of choice.Other traditional nongenetic biotechnologies include the fermentation of microorganisms to produce wine, beer, and cheese. Industry also uses microorganisms to produce various products such as enzymes for use in laundry detergents.

In an effort to find microorganisms that produce large amounts of enzymes, scientists sometimes treat a batch of organisms with radiation or chemicals to randomly produce genetic alternations. The process, called mutagenesis, produces numerous genetic changes in the bacteria, among which might be a few that produce more of the desired product.

Many new biotechnologies do not involve modifications of traits passed on to the next generation. A good example is monoclonal antibodies (highly specific preparations of antibodies that bind to a single site on a protein), which have many diagnostic applications, including home pregnancy testing kits.

Many biotechnology companies are engaged in these sophisticated, but noncontroversial, technologies. By contrast, mammalian cloning is a new biotechnology that does not involve gene modification, but is nevertheless highly controversial. Cloning reproduces adult mammals by transplanting a nucleus from adult cells into an egg from which the nucleus has been removed and allowing the egg to develop in a surrogate manner. The resulting individuals are as similar to the adults from which the nuclei were taken as identical twins are to one another. Although this procedure has profound implications for human reproduction, it does not modify specific traits of an individual, but rather transfers a whole nucleus containing a complete set of genetic information.

The new technology that can affect future generations is genetic engineering, a technology based on the artificial manipulation and transfer of genetic material. This technology can move genes and the traits they dictate across natural boundaries—from one type of plant to another, from one type of animal to another, and even from a plant to an animal or an animal to a plant. Cells modified by these techniques pass the new genes and traits on to their offspring. Genetic engineering can apply to any kind of living organism from microorganisms to humans.Genetic engineering can be applied to humans to replace or supplement defective genes. Where engineering is intended to cure disease, it is called gene therapy.

Potential applications that are not related to disease, such as the modification of traits like height, are sometimes called genetic enhancement. Currently, most genetic engineering of humans is done on nonreproductive or somatic cells, like those from bone marrow. The effects of this somatic cell gene therapy are confined to the treated individual. By contrast, germ line gene therapy would modify reproductive cells, so that the modification could be passed on to future generations.

Genetic Engineering Techniques

There are a number of techniques for moving genes artificially into recipient organisms. The oldest of these is called recombinant DNA, a technique that relies on biological vectors like plasmids or viruses. Other newer gene transfer techniques are electro—and chemical poration, microinjection, and bioballistics.

Recombinant DNA

Recombinant DNA techniques use biological vectors like plasmids and viruses to carry foreign genes into cells. Plasmids are small circular pieces of genetic material found in bacteria that have the ability to cross species boundaries. The circles can be broken and new genetic material added to them. Plasmids augmented with new genetic material can move across microbial cell boundaries and place the new genetic material next to the bacterium's own genes. Often the bacteria will take up the gene and begin to produce the protein for which the gene codes. Where the new gene codes for insulin, for example, the bacterium will begin to produce insulin along with its other gene products. A large vat of bacteria engineered to produce insulin can then become a sort of pharmaceutical factory.

Viruses can also act as vectors in genetic engineering. Viruses are infectious particles that contain genetic material to which a new gene can be added. The virus can carry the new gene into a recipient cell in the process of infecting that cell. The virus can also be disabled so that while it can carry a new gene into a cell, it cannot redirect the cell's genetic machines to make thousands of copies of itself.

Microinjection

Other methods do not rely on biological vectors like plasmids and viruses. One of these is called microinjection and involves simply injecting genetic material containing the new gene into the recipient cell. Where the cell is large enough, as many plant and animal cells are, the injection can be done with a fine—tipped glass needle. Somehow the injected genes find the host cell genes and incorporate themselves among them.

Electro-and Chemical Poration

Other methods for direct gene transfer involve creating pores or holes in the cell membrane to allow entry of the new genes. This can be done by bathing cells in solutions of special chemicals—so—called chemical poration—or subjecting cells to a weak electric current—so—called electroporation.

Bioballistics

Finally, there are so-called projectile methods that use metal slivers to deliver the genetic material to the interior of the cell. The small slivers (much smaller than the diameter of the target cell) are coated with genetic material. One projectile method, called bioballistics, propels the coated slivers into the cell using a shot gun. A perforated metal plate stops the shell cartridge, but allows the slivers to pass through and into the living cells on the other side. Once in the cell, the genetic material is transported to the nucleus where it is incorporated among the host genes.

Alternatives to Genetic Engineering

Alternatives can be complicated and elusive, and much depends on the goals envisioned. When the goals include reducing dependence on pesticides and herbicides, there are clearly alternatives to many biotechnology products. Many of these alternatives are not other products, but instead the systems and methods of sustainable agriculture.

A good example is crop rotation, which keeps pests under control elegantly by depriving them of the continuous food supply they need to build up large populations. Crop rotation has many advantages. It controls a broad variety of pests rather than just one or two. It does not select for resistance genes, as do chemical toxins or genetically engineered crops. And it does not result in ongoing pollution of air or water. As a pest-control strategy, crop rotation is far preferable to both chemical insecticides and genetically engineered crops.

Unfortunately, because it involves processes and not products, there is no industrial constituency to develop and support crop rotation as there is for the products of biotechnology. Conversion from industrial agriculture to sustainable systems that depend less on chemicals would eliminate the need for many of the currently projected products of biotechnology. This is not to say that there is no place for genetically engineered crops in sustainable systems; there may well be. But before such crops are introduced to sustainable agriculture systems, those systems must be more fully developed than they currently are.

The specific products engineered for sustainable agriculture would be different from those that are being developed to fit into industrial agricultural systems and their development should probably await the wider adoption of such systems. Agricultural biotechnology, as it is currently developing, is not particularly fruitful in the quest for a sustainable agriculture. Sustainable agriculture solves problems by understanding and adjusting the elements of the system to achieve its goals rather than by developing new products that must be purchased.

Agricultural biotechnology, by contrast, is basically an input industry, developing products, often expensive products. In sustainable agriculture, new products are less important than new knowledge and new ways of manipulating agricultural ecosystems.

Benefits of Genetic Engineering

Genetically engineered organisms have many potential applications in agriculture, including novel foods, pesticides, and animal drugs. Certainly, the companies that develop and market them believe they have benefits, and commercial success is one gauge of the need for and usefulness of a product. But the inquiry must go deeper. Society must consider whether these products are needed and whether better alternatives exist for meeting those needs. Whether such products deliver genuine benefits depends on the goals of agricultural and food systems, and the alternatives available for meeting those goals.

Today, in a world with abundant food, more than 700 million people are chronically undernourished. Over the next 20 years, the world's population will probably double. The global food supply would need to double just to stay even, but to triple for the larger population to be fed adequately. Limits in arable land and productivity and are employing practices that are destroying the soil's capacity to produce food.

But biotechnology, if by this we mean crops engineered to contain new genes, is not essential. It could play a minor and useful role in developing new agricultural products, but other factors—including other kinds of breeding technologies—will be much more important than transgenic crops in determining whether we meet this challenge. It would be a tragedy if other necessary actions were not taken because of a mistaken belief that genetic engineering is some sort of a panacea for hunger.

Solution to the World's Food Crisis

There are many reasons for the current and projected food crisis. Among the most important are lack of income to buy food, lack of infrastructure like roads to get products to market, trade policies that disadvantage farmers in the developing world, lack of inputs such as fertiliser, lack of information, and low-yield farming practices. More productive crops will do little to alleviate hunger if deficiencies in those areas are not addressed as well.

Where more productive crops are needed, there is little reason to believe that genetic engineering will be better than other technologies—in particular, sophisticated traditional breeding—at producing higher yielding crops. Many technologies can increase the yields of crops. These include traditional breeding, production of hybrids, so-called marker-assisted breeding (a sophisticated way of enhancing traditional breeding by knowing which plant cultivars carry which trait), and tissue culture methods for propagating virus-free root stocks. All of these could help improve the productivity of crops in the developing world, but currently only limited resources are available for applying them there.

So far, there no reason to believe that genetic engineering would be markedly better than these more traditional technologies in improving crops. Early "gene dreams" were of nitrogen-fixing crops, higher intrinsic yield, and drought tolerance. But so far none of these seems realistic because most involve complex multigene traits. For the most part, genetically engineered crops are limited to one or two gene transfers and have relative few applications of use to hungry people.

Those that are of use, such as insect resistance and virus tolerance, do not increase intrinsic yield and vary in effectiveness. In addition, they appear to be short lived due to the almost certain evolution of resistant pests. Currently, there is no reason to believe that the limited resources for agricultural development would be better spent on producing genetically engineered crops rather than on applying breeding technologies.

Most genetic engineering in agriculture is being done by large transnational corporations that need to sell their products at premium prices to cover the cost of research. These companies are developing products for farmers in rich countries who can afford to pay high prices for seed. Such farmers are interested in field crops like corn, soybeans, and cotton and fruits like tomatoes and cantaloupes.

And that is what the agricultural biotechnology industry is providing. In many cases, genetically engineered fruits are sold at premium prices and seeds are sold with an added technology fee to cover the costs of research. These products are of virtually no value to hungry farmers in Africa, who cannot afford the products of traditional technology, much less these expensive genetically engineered products. In addition, these products are often inappropriate for the developing world because, among other things, they require large amounts of fertilisers, pesticides, and water.

In sum, more productive crops are only part of the solution to the world hunger problem and transgenic crops are not uniquely capable of increasing food production. While some genetically engineered crops will undoubtedly prove useful, there is no reason at this time to invest huge sums in them, especially at the expense of traditional breeding.

Increase the Food Supply

Growing appreciation of environmentally destructive impacts has led to a renewed interest in agroforestry, intercropping, mixed crop-livestock operations as systems that can increase production with minimal chemical fertilisers and pesticides and a high degree of environmental protection. Much can be done to promote the sustainable intensification of agricultural production. Most of it should be done in developing countries to enable people to feed themselves so that they do not become dependent on commodities from abroad. All of it depends on local climates, cultures, and economic conditions.

Rice farmers in Southeast Asia, for example, are in a far different situation from farmers living at the edge of the Sahara desert. Among the many research areas important for increasing production are the efficient use of irrigation water, crop improvement through traditional plant breeding, and new ways to manage crop-pest interactions, such as integrated pest management.

Improvements in other parts of the agricultural system are also essential. These include building and maintaining roads so that farmers can get their crops to market, organising cooperatives so that farmers can purchase equipment and fertiliser, and reducing post-harvest losses of crops. Finally, meeting the world food crisis will require changes outside of agriculture like improving the incomes of the poor through microenterprises and shifting the diet of the rich away from excessive dependence on grain-fed livestock. Growing corn to feed cows and chickens is a much less efficient use of limited arable land than growing corn for humans to eat directly.

Risks of Genetic Engineering

Many previous technologies have proved to have adverse effects unexpected by their developers. Turned out to accumulate in fish and thin the shells of fish-eating birds like eagles and ospreys. And chlorofluorocarbons turned out to float into the upper atmosphere and destroy ozone, a chemical that shields the earth from dangerous radiation.

Being able to answer it depends on understanding complex biological and ecological systems. So far, scientists know of no generic harms associated with genetically engineered organisms. For example, it is not true that all genetically engineered foods are toxic or that all released engineered organisms are likely to proliferate in the environment. But specific engineered organisms may be harmful by virtue of the novel gene combinations they possess. This means that the risks of genetically engineered organisms must be assessed case by case and that these risks can differ greatly from one gene-organism combination to another.

So far, scientists have identified a number of ways in which genetically engineered organisms could potentially adversely impact both human health and the environment. Once the potential harms are identified, the question becomes how likely are they to occur. The recognition of this possibility does not by itself justify stopping the technology, but does put a substantial burden on those who wish to go forward to demonstrate benefits.

Potential Harms to Health

Here are the some examples of the potential adverse effects of genetically engineered organisms may have on human health. Most of these examples are associated with the growth and consumption of genetically engineered crops. Different risks would be associated with genetically engineered animals and, like the risks associated with plants, would depend largely on the new traits introduced into the organism.

Transgenic crops could bring new allergens into foods that sensitive individuals would not know to avoid. An example is transferring the gene for one of the many allergenic proteins found in milk into vegetables like carrots. Mothers who know to avoid giving their sensitive children milk would not know to avoid giving them transgenic carrots containing milk proteins. The problem is unique to genetic engineering because it alone can transfer proteins across species boundaries into completely unrelated organisms.

Genetic engineering routinely moves proteins into the food supply from organisms that have never been consumed as foods. Some of those proteins could be food allergens, since virtually all known food allergens are proteins. Scientists have limited ability to predict whether a particular protein will be a food allergen, if consumed by humans. The only sure way to determine whether protein will be an allergen is through experience. Thus importing proteins, particularly from nonfood sources, is a gamble with respect to their allergenicity.

Antibiotic Resistance

Genetic engineering often uses genes for antibiotic resistance as "selectable markers." Early in the engineering process, these markers help select cells that have taken up foreign genes. Although they have no further use, the genes continue to be expressed in plant tissues. Most genetically engineered plant foods carry fully functioning antibiotic-resistance genes.

The presence of antibiotic-resistance genes in foods could have two harmful effects. First, eating these foods could reduce the effectiveness of antibiotics to fight disease when these antibiotics are taken with meals. Antibiotic-resistance genes produce enzymes that can degrade antibiotics. If a tomato with an antibiotic-resistance gene is eaten at the same time as an antibiotic, it could destroy the antibiotic in the stomach.

Second, the resistance genes could be transferred to human or animal pathogens, making them impervious to antibiotics. If transfer were to occur, it could aggravate the already serious health problem of antibiotic-resistant disease organisms. Although unmediated transfers of genetic material from plants to bacteria are highly unlikely, any possibility that they may occur requires careful scrutiny in light of the seriousness of antibiotic resistance.

In addition, the widespread presence of antibiotic-resistance genes in engineered food suggests that as the number of genetically engineered products grows, the effects of antibiotic resistance should be analysed cumulatively across the food supply. Many organisms have the ability to produce toxic substances. For plants, such substances help to defend stationary organisms from the many predators in their environment. In some cases, plants contain inactive pathways leading to toxic substances.

Addition of new genetic material through genetic engineering could reactivate these inactive pathways or otherwise increase the levels of toxic substances within the plants. This could happen, for example, if the on/off signals associated with the introduced gene were located on the genome in places where they could turn on the previously inactive genes.

Concentration of Toxic Metals

Some of the new genes being added to crops can remove heavy metals like mercury from the soil and concentrate them in the plant tissue. The purpose of creating such crops is to make possible the use of municipal sludge as fertiliser. Sludge contains useful plant nutrients, but often cannot be used as fertiliser because it is contaminated with toxic heavy metals. The idea is to engineer plants to remove and sequester those metals in inedible parts of plants. In a tomato, for example, the metals would be sequestered in the roots; in potatoes in the leaves.

Turning on the genes in only some parts of the plants requires the use of genetic on/off switches that turn on only in specific tissues, like leaves. Such products pose risks of contaminating foods with high levels of toxic metals if the on/off switches are not completely turned off in edible tissues. There are also environmental risks associated with the handling and disposal of the metal-contaminated parts of plants after harvesting.

Although for the most part health risks are the result of the genetic material newly added to organisms, it is also possible for the removal of genes and gene products to cause problems. For example, genetic engineering might be used to produce decaffeinated coffee beans by deleting or turning off genes associated with caffeine production. But caffeine helps protect coffee beans against fungi. Beans that are unable to produce caffeine might be coated with fungi, which can produce toxins. Fungal toxins, such as aflatoxin, are potent human toxins that can remain active through processes of food preparation.

As with any new technology, the full set of risks associated with genetic engineering have almost certainly not been identified. The ability to imagine what might go wrong with a technology is limited by the currently incomplete understanding of physiology, genetics, and nutrition.

Environmental Harms

One way of thinking generally about the environmental harm that genetically engineered plants might do is to consider that they might become weeds. Here, weeds means all plants in places where humans do not want them. The term covers everything from Johnson grass choking crops in fields to kudzu blanketing trees to melaleuca trees invading the Everglades. In each case, the plants are growing unaided by humans in places where they are having unwanted effects. In agriculture, weeds can severely inhibit crop yield. In unmanaged environments, like the Everglades, invading trees can displace natural flora and upset whole ecosystems.

Some weeds result from the accidental introduction of alien plants, but many were the result of purposeful introductions for agricultural and horticultural purposes. Some of the plants intentionally introduced into the United States that have become serious weeds are Johnson grass, multiflora rose, and kudzu. A new combination of traits produced as a result of genetic engineering might enable crops to thrive unaided in the environment in circumstances where they would then be considered new or worse weeds. One example would be a rice plant engineered to be salt-tolerant that escaped cultivation and invaded nearby marine estuaries.

Gene Transfer

Novel genes placed in crops will not necessarily stay in agricultural fields. If relatives of the altered crops are growing near the field, the new gene can easily move via pollen into those plants. The new traits might confer on wild or weedy relatives of crop plants the ability to thrive in unwanted places, making them weeds. For example, a gene changing the oil composition of a crop might move into nearby weedy relatives in which the new oil composition would enable the seeds to survive the winter. Overwintering might allow the plant to become a weed or might intensify weedy properties it already possesses.

Crops genetically engineered to be resistant to chemical herbicides are tightly linked to the use of particular chemical pesticides. Adoption of these crops could therefore lead to changes in the mix of chemical herbicides used across the country. To the extent that chemical herbicides differ in their environmental toxicity, these changing patterns could result in greater levels of environmental harm overall. In addition, widespread use of herbicide-tolerant crops could lead to the rapid evolution of resistance to herbicides in weeds, either as a result of increased exposure to the herbicide or as a result of the transfer of the herbicide trait to weedy relatives of crops. Again, since herbicides differ in their environmental harm, loss of some herbicides may be detrimental to the environment overall.

Susceptible to Pesticides

Many insects contain genes that render them susceptible to

pesticides. Often these susceptibility genes predominate in natural populations of insects. These genes are a valuable natural resource because they allow pesticides to remain as effective pest-control tools. The more benign the pesticide, the more valuable the genes that make pests susceptible to it.

Certain genetically engineered crops threaten the continued susceptibility of pests to one of nature's most valuable pesticides: the Bacillus thuringiensis or Bt toxin. These "Bt crops" are genetically engineered to contain a gene for the Bt toxin. Because the crops produce the toxin in most plant tissues throughout the life cycle of the plant, pests are constantly exposed to it. This continuous exposure selects for the rare resistance genes in the pest population and in time will render the Bt pesticide useless, unless specific measures are instituted to avoid the development of such resistance.

Consequences of Foreign Genes

Addition of foreign genes to plants could also have serious consequences for wildlife in a number of circumstances. For example, engineering crop plants, such as tobacco or rice, to produce plastics or pharmaceuticals could endanger mice or deer who consume crop debris left in the fields after harvesting. Fish that have been engineered to contain metal-sequestering proteins (such fish have been suggested as living pollution clean-up devices) could be harmful if consumed by other fish or raccoons.

One of the most common applications of genetic engineering is the production of virus-tolerant crops. Such crops are produced by engineering components of viruses into the plant genomes. For reasons not well understood, plants producing viral components on their own are resistant to subsequent infection by those viruses. Such plants, however, pose other risks of creating new or worse viruses through two mechanisms: recombination and transcapsidation. Recombination can occur between the plant-produced viral genes and closely related genes of incoming viruses. Such recombination may produce viruses that can infect a wider range of hosts or that may be more virulent than the parent viruses.

Transcapsidation involves the encapsulation of the genetic material of one virus by the plant-produced viral proteins. Such hybrid viruses could transfer viral genetic material to a new host plant that it could not otherwise infect. Except in rare circumstances, this would be a one-time-only effect, because the viral genetic material carries no genes for the foreign proteins within which it was encapsulated and would not be able to produce a second generation of hybrid viruses.

Risks must be assessed case by case as new applications of genetic engineering are introduced. In some circumstances, it is possible to assess risks with great confidence. For example, it is vanishingly unlikely that genetically engineered palm trees will thrive in the Arctic regardless of what genes have been added. But for many potential harms, the answers are far less certain.

Risk assessments can be complicated. Because even rigorous assessments involve numerous assumptions and judgment calls, they are often controversial when they are used to support particular government decisions. For example, the approval of the first genetically engineered squash by the United States Department of Agriculture involved a controversial risk assessment.

Under the current US regulatory framework for biotechnology, some sort of risk assessment is routinely produced before decisions to allow commercialisation of products under the Federal Plant Pest Act; the Federal Insecticide, Fungicide, and Rodenticide Act (FIFRA); and the Toxic Substances Control Act (TSCA). In the case of the Plant Pest Act, risk assessments are done according to the procedure specified by the National Environmental Policy Act (NEPA). Under NEPA, risk assessments could lead to full-blown environmental impact statements, but so far all evaluations of engineered agricultural organisms have led to the legal conclusion that no environmental impact statement is needed.

For the most part, risk assessments are done by scientists and policymakers in the relevant agencies (USDA or EPA) with information provided by the companies seeking the approvals. A risk assessment for a microbial pesticide, for example, would be substantially different from a risk assessment for genetically engineered salmon. Like all efforts at risk evaluation, risk assessments done for regulation depend on the base of scientific knowledge for generation of list of possible harms to be assessed.

5

PLANT GENOMICS

The availability of whole-genome sequences as well as other genomic resources has allowed us to extend the comparative method to encompass the evolution of genome structure and function. More than just an isolated field, the past few years have witnessed the emergence of comparative genomics as a tool to address questions in diverse areas of biological research.

Characterisation of genomes, including whole genome sequences, has traditionally revealed numerous species-specific details, including genome size, gene number, patterns of sequence duplication, a catalogue of transposable elements, and syntenic relationships. At present, however, continue to know very little about the evolutionary dynamics of changes in genome structure and their consequences on gene content and evolution.

As more whole-genome sequences at both intraspecific and interspecific levels become available, we will be in a position to address several key issues surrounding the evolution of genome architectures. These include the extents and rates of change in genome structure and size, patterns of large-scale genome duplications (including polyploidisation), the dynamics of the origins and extinctions of genes, the role of selection acting on large-scale variation in genome structure and organisation, the evolutionary forces that determine transposable element activity and number and the functional consequences of these mobile elements, and the extent and impact of epigenetic markings on genome and organismal evolution.

Genome Function Evolution

Genomic tools have provided a boon for researchers seeking to understand the functional roles of genes and their evolutionary histories. Especially useful has been the appearance of genome-based methods to identify genomic regions of functional importance. The availability of intraspecific whole-genome sequences can reveal single nucleotide polymorphism genomic regions with markedly low or high levels, possible indicators of positive or balancing selection, both of which are signatures of adaptive evolution. The mark of selection on candidate loci identified in this manner can then be verified by sampling more individuals within the species.

The occurrence of intraspecific variation for phenotypic traits of interest also permits the identification of genes responsible for these traits by searching for associations among naturally occurring genome-wide polymorphisms using linkage disequilibrium mapping, a technique whose viability is SNP currently being actively investigated in plants. When candidate genes are already known, genomic variation data can provide the necessary controls for phenotypic association studies. Over the last several years, we have begun to realise that the products of genes are embedded in large-scale interaction networks that represent integrated functional units at the molecular genetic level.

Thus, to understand the evolution of function, it becomes necessary to understand the evolutionary dynamics of molecular genetic networks. Expression profiling with microarrays combined with EST and genome sequence data offer a viable way to do so, allowing us to identify the interacting candidate genes of genetic networks and examine how the patterns of interactions change across species. What is lacking is a solid theoretical framework for network structure and evolution, and work in this exciting area should be forthcoming.

Population Genomics and Phylogenomics

A key tenet of evolutionary genetics is that natural selection affects single genes or gene regions, but population processes, such as gene flow, range expansion, or bottlenecks, leave their imprint on all genes in the genome. Data for genome-wide polymorphisms for individuals of a species can now be easily obtained with high-throughput methods and is not limited to organisms with sequenced genomes. We are now faced with unprecedented amounts of genome information with which to characterise population history and structure.

Quantification of levels of intraspecific genome variation also aids in the identification of loci under selection, which exhibit patterns of variation divergent from those in the rest of the genome. As the level of structural variation in intraspecific genomes becomes more apparent, theory will have to be developed to describe genomic phenomena, such as the evolutionary dynamics of gene families and transposable elements, in a population context. A consequence of the proliferation of genome studies has been the documentation of patterns of genome variation between species, information that can be used to assist in the construction of the Tree of Life, including plants.

Given the evolutionary distances between many organisms targeted by whole-genome and EST sequencing studies, such data is more likely to aid in the characterisation of deep phylogenetic nodes of plant phylogeny. Moreover, the extensive genome coverage of the data can also be used to explore molecular clocks along phylogenetic branches as well as to understand the genomic changes characteristic of major evolutionary lineages. This will allow us to advance toward a central goal of plant biology: the reconstruction of the evolutionary history of all extant plant groups.

Functional Regions in Genomes

Identification of functional regions in genomes can be carried out by searching for conservation among genome sequences, as functional regions are believed to be under stabilising selection and should be preferentially conserved over evolutionary time. This approach has been used successfully in the annotation of animal and yeast genomes, but has not yet been used extensively in plants.

Both coding and regulatory regions can be identified by locating genomic areas under purifying selection, although regulatory regions tend to be less conserved and pose greater challenges to the comparative genomics approach. Comparisons at varying levels of evolutionary divergence are likely to reveal functional regions characteristic of different plant groups; even intraspecific genomic approaches have been shown to be useful in predicting functional sequence motifs. The reliability and usefulness of comparative genomics for genome annotation will depend on the continuous improvement of predicting algorithms, as well as our improving characterisation of the varying neutral evolutionary rates across sequenced genomes.

Comparisons among multiple species, although computationally intense, have also been shown to be a powerful method in the prediction of functional genomic regions. The success of such approaches for plant genome annotations will hinge on the completion of sequenced genomes for plants of all major evolutionary lineages.

Comparative genomics has proven an invaluable approach to understanding biology, not only for dissecting patterns and processes of genome evolution but also in revealing aspects of gene function. The rapid advances in technology, both for sequencing and for determining expression and interaction patterns, will continue to propel this area in the future. Although it is as yet unreasonable to expect that everybody's favourite organism will be sequenced to completion, the plant research community as a whole would benefit from candidate genomes chosen within a reasonable phylogenetic framework.

Ideally, this would include candidates from non-seed plant lineages, gymnosperms, and major angiosperm lineages, and steps for at least comparative EST analysis in this regard are under way. Maximum benefit could be derived from applying Bennetzen's suggestion to sequence pairs of organisms for each selected lineage. Intraspecific genome comparisons will primarily rely on resequencing techniques in the near future, though the advent of chip-based genomic array techniques as well as new methods will make it easier to acquire large genome coverage for individuals within populations or species. These tools, coupled with functional genomics approaches, may provide crucial insights into how genomes evolve in structure and function, and also permit us to link genome structure with organismal biology.

Genome Scrambling

Internationally, safety regulations of transgenic (genetically modified or GM) crop plants focus primarily on the potential hazards of specific transgenes and their products. This emphasis on the transgene and its product is a feature of the case-by-case approach to risk assessment. The case-by-case approach effectively assumes that plant transformation methods carry no inherent risk.

Nevertheless, current crop plant transformation methods typically require tissue culture (i.e. regeneration of an intact plant from a single cell that has been treated with hormones and antibiotics and forced to undergo abnormal developmental changes) and either infection with a pathogenic organism (A. tumefaciens) or bombardment with tungsten particles. It would therefore not be surprising if plant transformation resulted in significant genetic consequences which were unrelated to the nature of the specific transgene. Indeed, both tissue culture and transgene insertion have been used as mutagenic agents.

Transformation-induced Mutations: In theory, plant transformation could result in exact insertion of a single transgene without further genomic disruption. In practice, this rarely, if ever, occurs. In addition to the transgene, each transformed plant genome contains a unique spectrum of mutations resulting from (a) tissue culture procedures, (b) gene transfer methods such as Agrobacterium-mediated or particle bombardment transfer, (c) transgene insertion and (d) superfluous DNA insertion. These transformation- induced mutations can be separated into two types: those introduced at the site of transgene insertion, which refer to as insertion-site mutations and those introduced at other random locations, which refer to as genome-wide mutations.

Insertion-site mutations: Remarkably little is known about the mutations created in crop plants at the site of transgene insertion. This is true both for transgene insertion via Agrobacterium-mediated transformation and for particle bombardment. This lack of understanding is caused in part by a lack of large-scale systematic studies of insertion-site mutations. Additionally, much of the available data comes from research on a non-crop plant, Arabidopsis thaliana, and it is not clear whether such results apply to crop plants.

Agrobacterium-mediated transformation: Agrobacterium-mediated transformation has been used to create commercial cultivars for over 10 years and is known to create insertion-site mutations. However, there has been only one large-scale study of the mutations created at insertion events containing single T-DNA inserts. Most of the T-DNA insertions resulted in small (1-100 base pair) deletions of plant genomic sequences at the insertion-site. However, for a significant number (24/112) there was evidence for largescale rearrangement of plant genomic DNA at the insertion site. Two of these insertion events contained chromosomal translocations.

The rest had rearrangements which were not fully characterised. It is known, however, that rearrangements of genomic DNA at T-DNA insertion sites can be very substantial. A 78Kbp deletion (removing 13 genes) is the largest recorded for T-DNA insertion and other researchers have reported duplication and translocation of a segment of DNA at least 40 Kbp in size. Superfluous DNA insertion is also a common feature of T-DNA insertion-sites. Vast majority of T-DNA insertion events include small or large genomic DNA disruptions and insertions of superfluous DNA.

Particle bombardment transformation: Particle bombardment has also been used to create numerous commercial cultivars. Although it can result in large scale genomic disruption, there are few studies detailing the insertion-site mutations resulting from particle bombardment. Furthermore, there have been no large-scale systematic studies of such mutations. Most of the particle bombardment insertion events that are described in the scientific literature are extremely complex.

Multiple copies of delivered DNA are often interspersed with small or large fragments of plant genomic DNA.. Without the use of PCR and DNA sequencing, analyses of insertion-site mutations are likely to be incomplete. Researchers have found only two particle bombardment studies where PCR and DNA sequence analyses were used to characterise mutations created at single-copy insertion events which had been isolated from intact plants.

Some data describing particle bombardment insertion events are extremely limited. However, some other data suggest that transgene integration at particle bombardment insertion events is always accompanied by substantial genomic disruption and superfluous DNA insertion. Southern blot analysis is insufficient to identify all insertion-site mutations: Another limitation to the understanding of insertion-site mutations is that Southern blot hybridisation is the technique most commonly used to analyse transgene insertion events for both research and regulatory purposes.

Analysis of transgene insertion-sites by other techniques such as FISH, PCR or DNA sequencing indicates that Southern blot analysis is not sufficient to reliably determine either the presence of superfluous DNA or the extent of genomic disruption at the transgene insertion-site. For example, Mehlo et al. used both PCR and Southern Blot analysis to analyse particle bombardment insertion events and concluded that Southern blotting was useful only in detecting large-scale features of the transgene insertion-site. In another study, fiber-FISH techniques were used to analyse a particle bombardment insertion event which was predicted by Southern blotting to contain tandem repeats of a transgene. Their analysis revealed that there were actually 3-10 Kbp of chromosomal DNA between most of the repeats. This suggests that, in some cases, Southern blot analysis is inadequate for identifying even large-scale rearrangements.

All reports lead to draw various conclusions. Firstly, that analysis of transgenic lines based solely or primarily on Southern blot data can miss many of the mutations present at insertion-sites. Thus, the plant genome is probably more disrupted by transgene insertion than commonly supposed. Secondly, that, as almost all commercial approvals of transgenic events or cultivars are based primarily on Southern blot analysis of transgene insertion, it is likely that most transgenic events currently approved for commercial use harbour unreported large and small-scale transgene insertion-site mutations.

Genome-wide mutations: There are 5 studies in which researchers have attempted to quantify the number of mutations introduced during plant transformation. These researchers used DNA polymorphism analysis (based on RFLP, AFLP and other PCR techniques) to compare the genomes of transformed plants to the genomes of non-transformed control plants.

Their results suggest that many hundreds or thousands of such genome-wide mutations are likely to be present in plants transformed using typical plant transformation methods, especially those involving the use of plant tissue culture techniques. In one study, Labra et al. estimated that the "genomic similarity value" of control plants was 100%, but only 96- 98% for the transgenic plants. In other words, very extensive genetic mutation had resulted from the plant transformation procedures. Even though the numbers of mutations found in these studies were high, the nature of the analytical techniques used in these experiments suggests that these figures may underestimate the extent of mutation to the plant genome.

Also, such studies do not address the nature of these mutations, such as whether they are small scale or large-scale genomic changes and whether they occur in functional regions of the genome. Depending on the extent of outcrossing or backcrossing undergone by the primary transformant, many and sometimes all of the mutations created in the primary transformant are likely to be retained in commercialised cultivars. Even where backcrossing has been extensive, genome-wide mutations genetically linked to the transgene insertion-site probably remain in the commercial cultivar.

Genome-wide mutations have been found in all transformed plants examined and such mutations have been shown to be heritable. However, current safety regulations do not require any testing or analysis of genome-wide mutations in commercial cultivars.

Significance of transformation-induced mutations: Insertion-site and genome-wide mutations can be hazardous if they occur in a functional region of plant DNA. Mutations in functional plant DNA, including gene coding sequences or regulatory sequences, may have implications for agronomic performance or environmental interactions or for animal or human health. For example, a transformation-induced mutation might disrupt a gene whose product is involved in nutrient biosynthesis, resulting in altered nutrient levels, or it might disrupt or alter a gene involved in the regulation or synthesis of compounds toxic to humans.

Disruption of a gene encoding a regulatory protein, such as a transcription factor, could result in the miss-expression of numerous other genes. Such biochemical changes would be unpredictable and difficult to identify even with extensive biochemical testing . Typically, only a few biochemical tests are required by regulators. Therefore, using current safety assessments, many of the harmful phenotypes which could arise from transformation-induced mutations would be unlikely to be identified prior to commercialisation.

Frequency of disruption of functional DNA by transformation-induced mutations: The limited amount of data available suggests that transgenes frequently insert into or near gene sequences. In the few plant species studied, DNA sequence analysis of T-DNA insertion-sites suggests that approximately 35-58% of transgene insertions disrupt plant gene sequences. Similar studies of transgenes delivered via particle bombardment have never been conducted. Despite its importance for safety assessment, it is usually not clear whether transgenes in commercial lines have inserted into or near gene sequences.

Most applications submitted to the USDA requesting permission to commercialise a transgenic line provide neither the sequence of the genomic DNA flanking the inserted transgene nor a comparison with the original target-site sequence. An added difficulty in determining the significance of an insertion event is that it is currently not possible to know with certainty that a region of the genome is non-functional. The frequency with which genome-wide mutations disrupt functional DNA has never been specifically investigated.

However, the successful use of tissue culture to induce mutations for research and breeding purposes and the isolation, from populations of transformed plants, of mutant phenotypes which are not linked to a transgene insert both suggest that genome-wide mutations do frequently occur in functional DNA sequences. Even if no functional sequences are disrupted, transgene and superfluous DNA insertions are not necessarily harmless or inert. Promoter sequences may alter the expression of neighbouring genes, while bacterial chromosomal or plasmid sequences (bacterial origins of replication in particular) inserted adjacent to the transgene may enhance the probability of horizontal gene transfer.

6

GENE FLOW IN PLANTS

Interest in gene flow was largely restricted to a modest subset of evolutionary and population biologists, and to plant breeders, for whom it was important to interpret patterns of phylogeny, adaptation, and degree of contamination in seed production populations. With the advent of genetic engineering, and the biopolitical and legal controversies it has generated, gene flow has moved onto the public stage. "Genetic pollution" due to inadvertent mixing of seeds and pollen flow was the cause of "Tacogate," which caused Star-Link corn to be withdrawn from the market and the responsible company to pay compensatory damages in excess of US $100 million.

Labelling regulations for GMO crops that are under development in many countries appear to be requiring strict limits on the extent of transgene flow and crop mixing. The extent to which we understand, can predict, and can efficiently monitor gene flow may therefore be the most important biological determinant of whether transgenic crops are adopted and publicly accepted. Because of the extensive gene flow possible from trees, and their very limited history of domestication, it has long been known that genes from intensively bred or engineered trees are highly likely to enter wild populations unless very special measures-such as use of sterile trees-are taken to avoid it.

In the southern hemisphere, where some of the most intensive plantation forestry in the world occurs, most of the trees grown are exotics, thus greatly reducing concerns over "pollution" of wild populations. However, in many of these places the planted trees have feral, "naturalised" forms, and in some cases these have given rise to invasive weeds that, because of their size, can have substantial impacts on local ecosystems. Thus, gene dispersal of exotics may present even more of an ecological hazard than that from native trees-where there is often a large "buffer" provided by wild populations, and where the trees already occupy defined niches.

Although forestry crops are mostly not consumed as food, trees-even when they are farmed like row crops-tend to be considered symbols of durability and wildness by the public. And while it is true that people tend to worry more about the quality of their food than they do about the quality of their paper, it is also true that people are far more passionate about their trees than they are about soybean or broccoli crops. Perceived "genetic adulteration" via gene flow, even at a low level, may thus be a significant impediment to public acceptance of transgenic plantations.

Transgene flow, and means to predict its frequency and consequences, are therefore globally important issues for the future of plantation forestry. Unfortunately, we have a limited knowledge base for risk assessments. There have been very few studies of gene flow from conventional plantations to wild populations. Due to the long lifespans and generation times of trees, predictions of impacts must consider very large temporal and spatial scales-and thus embrace high levels of uncertainty about future ecological conditions. New studies, and new tools, will be needed to make rational inferences about possible impacts of transgenic plantations.

Early studies of gene flow in forest trees focused on the physical dispersal potential of pollen and seeds. As early as 1919, pollen grains of Pinus sylvestris had been observed far beyond the range of the species. Later studies confirmed long-distance movements for both pollen and seed. However, tracking of propagules provides estimates of potential gene flow; realised gene flow also depends on fertilization and establishment success. Factors such as pollen competition, phenological synchrony, and seed predation can result in levels of gene flow far below those predicted by propagule movement alone. Nonetheless, these studies demonstrated the astonishing mobility of pollen, and to a lesser extent, of light seeds.

The advent of genetic markers revolutionised methods of measuring gene flow. Over the last three decades, rates of migration have typically been inferred from the degree of genetic differentiation among populations as measured by the fixation index FST, or its many extensions and analogs. A common approach has been to estimate the fixation index from allele frequency data and then convert this to the mean number of migrants per generation (N_em) using the relationship ST

$$N_em \approx \frac{1 - F_{ST}}{4F_{ST}}$$

There are many methods for estimating the fixation index, each with strengths and weaknesses. However, the most common approach in forest trees has been to use allozyme allele frequencies to calculate GST, and derive N_em using Crow and Aoki's correction for the number of populations (or subpopulations) in the sample. Other approaches for measuring historical gene flow include the "rare allele method" and coalescent methods. The rare allele method is based on the approximately negative linear relationship between the logarithm of m Ne and the average frequency of private alleles in the demes of a genetically subdivided population. The large sample sizes needed for application of this method and the required availability of rare alleles have limited the use of this approach in tree population studies.

Gene coalescence-based methods are based on genealogical relationships among alleles. They appear to be superior to other methods when variation within populations exceeds variation among populations. They should therefore be highly suitable to forest trees. However, we are aware of no studies that have applied this method to any forest tree taxa. All such methods of measuring gene flow are 'indirect', because they apply genetic models (and underlying assumptions) to infer long-term levels of gene flow.

For example, the correlation between FST and $N_e m$ is often unreliable because the assumptions of the underlying population structure models are rarely met in real populations. Indirect measures reflect the complex interactions of all demographic parameters and evolutionary forces acting on a population, and the resulting gene flow estimates should be taken as long-term averages estimated over a large number of populations. Despite these shortcomings for estimating contemporary gene flow, indirect approaches have provided a number of valuable insights about historical forces shaping forest tree genetic structure. There have been a large number of studies of allozyme gene diversity, geographic structure, and gene flow among populations of forest trees, and a few generalisations have emerged:

1. Trees are characterised by higher genetic diversity and lower levels of differentiation compared to other plant groups. The interpopulation component of total gene diversity (based on the fixation index) of woody species rarely exceeds 10-15% (Table 1). This low differentiation suggests extensive gene flow among tree populations. However, some authors have hypothesised that the observed patterns of genetic variation may also be due to the long pre-reproductive phase of most tree species mitigating founder effects, and trees' long generation span retarding differentiation through genetic drift.

2. Wind-pollinated tree species typically have interpopulation differentiation levels of less than 10%, which translates to

more than 2 successfully established migrants per population in each generation. In species with large and continuous ranges, interpopulation differentiation is often below 3%. On the opposite extreme, species with small and fragmented populations can have interpopulation differentiation in the range of 15 to 30%. This observation suggests that even though long-distance pollen dispersal is possible, its effect may be insufficient for genetic homogenisation of spatially isolated populations.

Table 1. Neighborhood sizes and per-generation migration events estimated by two indirect methods for some tree genera.

Genus	GST	$N_e m$	$N_e m^*$	Na
Abies	0.063	3.72	8.71	260
Picea	0.055	4.30	—	133-260
Pseudotsuga	0.074	3.13	—	—
Pinus	0.065	3.60	1.51-21.83	158-534
Quercus	0.107	2.09	2.24-6.74	—
Populus	0.041	5.85	—	—
Eucalyptus	0.169	1.23	0.76-6.51	—

GST - averages

$N_e m$ - estimated based on the GST values

$N_e m^*$- estimated for some species using the method of Slatkin (1985); from Govindaraju (1989)

Na - neighborhood area in m² (area from which the parents of some central individual may be treated as randomly drawn) estimated for some species

3. Outcrossed animal-pollinated tree species have a detectably (but not significantly) higher degree of interpopulation genetic differentiation compared to wind-pollinated trees. The average interpopulation differentiation for animal-pollinated trees with mixed mating systems was 10% based on 37 data sets. As in wind-pollinated species, spatial distribution seemed to be a good predictor of the degree of differentiation.

Direct observations of gene dispersal obviate the need for tenuous assumptions about historical conditions. Instead, they provide short-term "immigration snapshots." Generally, direct methods require the genotyping of all potential parents in a population and estimation of the proportion of progeny that could not have been produced by within-population mating. One approach employs simple paternity exclusion, feasible where the maternal genotype can be readily determined. However, the low variability of allozyme markers greatly limits the ability to distinguish between local and immigrant genotypes.

To help overcome this problem, a number of methods employ maximum likelihood to either assign parentage, or to estimate mating parameters that provide the best fit to observed progeny genotypes. In the early 1990s, highly variable DNA-based markers began to become affordable for parentage analyses. The high genetic resolution provided by microsatellite and AFLP data allowed scientists to conduct paternity analyses based on genotypic exclusion with acceptable levels of discrimination, as well as to apply maximum likelihood assignments with higher confidence.

Seed Orchards

There have been a number of studies of genetic contamination of forestry "seed orchards." Seed orchards are plantations in which selected genotypes are placed to allow cross-pollination for production of seeds to be used for reforestation. Most of these orchards are within the range of native or planted populations, and distinct orchard blocks (that service distinct ecogeographic regions) are in many cases planted in proximity for management efficiency. Thus, there are often high levels of unwanted pollen immigration into the orchard blocks. This can result in substantial loss of genetic gain compared to expectations based on selection theory, and can compromise adaptation of seed orchard crops to their intended plantation environments.

The tracking of rare allozyme alleles has been used to measure orchard contamination, and to study mating patterns within seed orchards. However, such alleles are often difficult to identify, and force inferences to be based on a very limited number of genotypes. Instead, pollen contamination of seed orchards is usually estimated via simple paternity exclusion, adjusting the observed proportion of immigrants by the probability that an immigrant gamete will be distinguishable from the potential orchard gamete pool. Other statistical procedures have also been implemented. These studies have revealed great variation in pollen contamination, even when the same analytical approaches and tree species are considered (Table 2).

Table 2. Estimated pollen contamination in clonal seed orchards of forest trees.

Tree species	S	Di	b	m
Picea abies	13.2	none	—	0.55-0.81
		—	0.10-0.17	
Picea glauca	—	1000	—	0.01
Pseudotsuga menziesii	1.8-3.3	none	—	0.29-0.91
	20	500	—	0.11
Pinus sylvestris	22.9	2000	—	0.48
	3	1000	0.15	—
	6	500	0.38	—
	12.5	>100	—	0.72
Pinus taeda	2	100	—	0.36
Pinus maritima	—	none	0.36	—
Eucalyptus grandis x E. urophylla	7.4	400	0.14	—
	8	800	0.03	—

S - area in which all potential parents have been genotyped (ha)

Di - distance to the nearest population or tree of the same species (m)

b - observed proportion of immigrant pollen gametes

m - pollen contamination adjusted for the probability to distinguish local and migrant gametes

Nonetheless, it is clear that pollen contamination is often very large, commonly exceeding 40%, even when the closest stands of the same species are several hundred meters away. Less is known about pollen contamination in seed orchards of animal-pollinated tree species such as eucalypts compared to that of the wind-pollinated conifers. However, these studies also suggest that contamination can be considerable, implying that pollinators can travel long distances from outside orchards-as well as move pollen extensively within orchards.

Due to restrictions on marker resolution and the high cost of genotyping, spatially distinct forest stands are usually chosen for study, and parentage analysis then applied to a sample of progeny. Typically, seeds are collected from mother trees of known genotype, and paternity estimated by comparing inferred paternal genotypes with those of all potential fathers in the analysed population. Some researchers have also sampled seedlings and/or saplings, attempting to estimate both pollen- and seed-mediated gene flow. However, such analyses require considerably more exclusion power than paternity analysis.

Results from paternity analyses in wind-pollinated species generally agree with predictions from studies of genetic structure. In most cases, the frequency of immigrant pollinations was over 30%, but appeared rather low (6.5%) in a spatially isolated population of Pinus flexilis. Successful pollen traveled as far as 10 km. As with indirect methods, gene flow estimates based on parentage analysis in animalpollinated tree species were somewhat lower, but generally similar to immigration rates in windpollinated trees. These results support the notion that animals can be effective agents of long-distance pollen and seed dispersal.

However, there is also considerable variance in pollen immigration between species, even when isolation distances are similar, cautioning against broad generalisations. Although paternity analysis based on highly variable markers appears to be the most effective current method for measuring gene dispersal in ecological time, estimates can be greatly affected by the presence of null alleles, and the misgenotyping of complex microsatellite and AFLP phenotypes. Both of these types of errors will cause overestimates of gene flow.

DNA Markers

One of the limitations of parentage analysis using nuclear loci is that it is often difficult to distinguish the contributions from the different agents of gene flow. The predominant uniparental inheritance of organelle DNA, however, provides a means to differentiate pollen—from seed-mediated migration. Chloroplasts are inherited maternally in most angiosperms, and paternally inherited in most gymnosperms. Plant mitochondria are generally passed on maternally, except for some gymnosperms. The high amounts of length variability detected in some regions of the chloroplast genome facilitate its use for gene flow studies.

Chloroplast microsatellites can increase efficiency of paternity analysis, and aid the estimation of pollen contamination in conifer seed orchards. Applied in combination with nuclear loci, plastid DNA markers are also valuable for inferring historical levels of gene flow. For example, Ennos compared FST values calculated over nuclear and organelle DNA markers and concluded that pollen-mediated gene flow in wind-pollinated species may exceed seed-mediated gene flow by a factor of 18 to 68 (in the light-seeded pines), and up to 196-fold in the acorn-bearing oaks..

Analysis of the spatial structure of cpDNA enabled estimates of postglacial seed dispersal rates, and recolonisation routes, of European oaks. However, because of the lack of independence of polymorphic loci in organelle genomes, differences in mutation rates among genomes, and difficulties in assuming homology among repeat "alleles," organelle-based inferences should be treated cautiously.

Spatial Simulation Modelling

Extrapolation of short-term or historical gene flow observations to spatial and temporal scales that are relevant for management and ecological policy remains a major challenge. Because of the time and expense required for a typical parentage analysis study, only a limited number of populations and years can be examined. However, ecologically significant levels of establishment may occur only once or twice per generation (i.e., on a decadal scale), and in particular habitats.

An emerging solution is the use of spatial simulation models to extrapolate results of short-term gene flow studies with knowledge of ecological processes. They provide an extensible framework for integrating data from disparate demographic and genetic field studies with landscape-scale analyses of ecosystem dynamics. In addition, such models allow 'virtual experiments' through sensitivity analyses in which selected components of the system are manipulated to determine their importance in determining long-term outcomes.

7

CROP IMPROVEMENT

To provide enough food for the expanding population, we need to produce as much food in the next 20 years as we have in the last 10, 000 years. Increased crop yields, pest resistance and tolerance to environmental stresses have been the targets of plant breeders during the last few decades. There is also progress towards improving food quality, as well as quantity, by improving the nutritional benefits of food crops. For example, the levels of vitamins, unsaturated fats, antioxidants and anti-carcinogenic compounds in various plants are all being enhanced.

There is also a great demand for 'speciality crops' for non-food use. For example, crops that produce starch, such as maize and potato, and those that produce oil such as oil seed rape (canola) are being intensively researched to develop plants that make different types of these polymers. These novel compounds can be used to make plastics and detergents, which at present are made using fossil fuels as a starting point. As the world's supply of crude oil and natural gas dwindles, there is a huge demand for alternative, 'renewable' energy resources.

Transformed plants can be valuable 'bioreactors' to produce large quantities of plant-derived pharmaceuticals, as it is potentially possible to increase the amount of these chemicals made by some plants. In addition, new strains of plants that make vaccines against diseases such as cholera and hepatitis are currently in clinical trials. For example, bananas and potatoes have been developed to make a protein that mimics part of a virus. By eating the plant, our bodies make antibodies against the virus, eliminating the need for vaccinations using needles. This could be particularly significant in developing countries, where there is a need for trained nurses using refrigerated vaccines to perform injections.

Methods for Crop Improvement

The traditional method of crop improvement is by sexual hybridisation, making crosses between two plants with desirable characteristics. At fertilisation, the genetic information from each parent recombines, so characteristics from each parent are inherited by the offspring. However, many other features are also inherited, along with the desirable one. A breeding programme lasting for many years may be necessary to eliminate the unwanted characteristics, without losing the desired one(s).

Alternative methods to develop new crop varieties use chemicals or radiation to cause imprecise alterations to the plant's genetic information. These cause random changes to the DNA may, in some cases, be of agricultural benefit to the plant. One of the advantages of mutagenesis is that it does not depend on crossing two sexually compatible species. However, time—consuming and laborious analysis of the offspring of the mutagenised seed is necessary to find a plant with the desirable characteristics.

Genetic modification, or plant transformation, is the newest method of crop improvement to become available to plant breeders. Using this approach, a fragment of DNA carrying the genetic information for a particular characteristic can be identified in any species of plant, animal or microorganism, then isolated, copied, and introduced into an existing crop variety. While this approach is unlikely to completely replace the existing methods of plant breeding, it potentially provides a wider gene pool from which to develop novel crops varieties.

Plants Transformation

Plant transformation is the introduction of small fragments of DNA into the chromosomes of a host plant. Chemical tools act as molecular 'scissors' and 'glue', enabling DNA fragments to be cut and pasted together in specific combinations before being introduced into a plant. As well as the gene encoding the characteristic of interest, a 'marker' gene is also introduced, to allow identification of transformed plants. These marker genes have traditionally conferred resistance to antibiotics. However, with widespread concerns about increasing antibiotic resistance, these genes are increasingly being replaced by alternatives.

Agrobacterium

Agrobacterium is a naturally occurring bacterium that causes crown-gall disease on wounded plants. During the infection process, the bacterium transfers a piece of its own DNA into the plant chromosomes. By engineering these bacteria to carry DNA encoding the characteristic of interest, a new gene can be transferred into pieces of wounded plant tissue. The bacteria used are disabled to prevent them causing disease on the plant. The pieces of plant tissue are then regenerated into whole plants on agar, a type of nutrient jelly that contains growth substances enabling a piece of plant tissue to form roots and shoots and develop into a whole plant.

Gene Gun Fires

The gene gun fires tiny gold particles coated in DNA into a piece of plant tissue, such as the embryo. The particles enter the tissue at high speed, so that they penetrate the tough cell wall. The DNA becomes dispersed within the cell and, in some cases, becomes incorporated into the plant,s own DNA. A whole plant carrying the novel gene in all its cells, can then be regenerated from the embryo.

Electroporation

This method involves incubating individual plant cells in a

solution containing the DNA encoding the gene of interest. By passing an electric current through the solution, the plant cell membranes become permeable, and the DNA is taken up. Under suitable culture conditions, an intact plant can again be regenerated from the individual cells.

Traditional and New Crop Improvement Methods

The so-called 'traditional' methods of crop improvement are limited by the sexual compatibility of the plants involved. The most significant advance in crop improvement to arise from the 'new' methods of genetic modification is that, theoretically, a characteristic from any organism, of any species, can be introduced into a plant. Plant breeders therefore have access to a much wider gene pool than they have using crossing methods. For example, one of the genes introduced into plants to provide resistance to the herbicide RoundUp™ was originally isolated from bacteria.

An insecticidal toxin used as a crop spray was also first identified in bacteria Genetically modified cotton and maize expressing this toxin is currently being grown in the United States. The other major difference between 'old' and 'new' methods of crop improvement is the number of genes transferred to the offspring in each case. On average, plants contain approximately 80 000 genes which recombine during sexual hybridisation. The offspring may therefore inherit around 1 000 new genes as a result of this recombination. This is equivalent to a 0.0125% change in the genome.

By contrast, only one or two new genes are transferred during plant transformation. As this represents a 0.0025% change in the genetic information of the plant, it is argued that plant transformation provides a more precise approach to crop improvement than sexual hybridisation. At present, legislation prevents any gene of unknown function and sequence from being introduced into plants. This means that, prior to carrying out the transformation, the novel gene must be thoroughly characterised. It is therefore possible to predict and monitor the behaviour of the gene following its incorporation into the plant DNA. This is critical to enable the long term effects of plant transformation to be monitored.

Currently, there is no evidence that 'transgenes' behave any differently than the 'resident' genes present naturally in the plant. The position and behaviour of a transgene can be identified and analysed more easily than any of the 80 000 resident genes. It has been suggested that genetic modification could cause harmful toxins to be made by transformed plants. However, it is unclear whether it is due to the technique itself, or to the nature of the foreign gene. For example, the introduction of a gene that is known to encode a toxin in one organism will almost certainly have a similar effect when introduced into a different organism.

It was, therefore, not surprising that the highly publicised transgenic soybean containing a gene from Brazil nuts also elicited an allergic reaction among sensitive patients. The gene from Brazil nut had been well characterised and its product known to be allergenic, hence the extensive laboratory tests. This illustrates why rigorous characterisation of a gene is required before permitting its introduction into a novel species. It is also important to look at potentially toxic products that have arisen in plants as a result of 'traditional' methods of crop improvement. For example, potato varieties with increased pest resistance have continually been selected as giving higher crop yields.

These varieties contain high levels of natural pesticides, called glucoalkaloids. However, these compounds are toxic to animals, so could have harmful effects when eaten. This demonstrates that the nature of the novel feature should be open to question, rather than the method by which it is introduced. Practically speaking, one of the biggest differences between genetic modification and traditional plant breeding is that, while extensive restrictions are in place to limit the development and release of genetically modified varieties, those developed by sexual hybridisation and mutagenesis are under no restrictions.

A major concern surrounding the cultivation of genetically modified crops is the possibility of cross pollination between transgenic and related crops. While this is clearly possible for some species, it is important to remember that not all crop species have native wild relatives with which they are sexually compatible, meaning that the generation of 'superweeds' is not as likely as is often suggested. Also, many plants, such as carrots, are only allowed to flower for seed production, which means that of cross-pollination during normal commercial cultivation procedures is unlikely.

Morals and Ethics of Genetic Modification

Many ethical objections have been raised to plant transformation. One of the most common is that it is unnatural, and those that carry it out are 'playing God'. However, some supporters argue that it is the intent to make changes, rather than the method by which they are brought about, that is the real subject of the ethical debate. From this point of view, humans have been manipulating the 'natural' world for millennia.

Other concerns are based on the potential problems that could arise from introducing genes that are 'foreign' (i.e. from other species) into plants. Critics argue that these genes could not have become naturally incorporated into the plant genome, and could therefore be potentially hazardous. However, what makes one species a 'foreigner' to another? The DNA from all organisms on the planet is chemically identical, it is the sequence of the genetic information that differs.

Even so, there are often very close similarities between the gene sequences of totally unrelated species from different kingdoms. It might therefore be argued that all organisms descend from a common ancestor and, as such, are distantly related. This idea is particularly important when considering ethical or religious objections about the introduction of genes from certain animals into plants.

Techniques of Transgenic Crop Improvement

Molecular biology provides several techniques to assess gene function, and indeed most efforts have been applied to the development of novel strategies to demonstrate the expression and translation of a particular gene. Transcript analysis (transcriptomics) and protein profiling (proteomics) are the most advanced strategies to provide an answer to the question:- How does the plasticity of gene expression respond to any particular external or developmental stimulus?

The identification of genes that control economically important traits provides the basis for new progress in genetic improvement of crop species, complementing traditional methods based on assisted crosses. This issue has gained particular importance, since molecular biologists first used transgenic methods in crop improvement. Initially, single genes were introduced, but after a few generations some transgenics no longer expressed the acquired characteristic, thus indicating either a physical loss of the introduced gene or a loss in its expression.

Similarly, even when transgenic plants carried an exogenous gene designed to provide protection against a particular pathogen, they remained susceptible because the response was not monogenic, as was initially believed. It was therefore essential in these cases to more fully determine the plant response to a particular biotic or abiotic stress. The identification of candidate genes influencing any important trait can be approached through an analysis of complementary DNA (cDNA), copies of messenger RNA (mRNA).

Devoid of intronic and intergenic sequences, whose biological significance is still obscure, these mRNAs represent only a small percentage of the total genome (about 1-3% in eukaryotes). However, they do contain valuable information on gene activity since they correspond to the proteins expressed in a specific tissue and responsible for the identity of that tissue. The formal identification of genes proceeds first by sequencing the expressed sequences tags (ESTs), partial sequences from either end of the complementary cDNA, and then by cloning the complete gene.

Sequence information from ESTs can be used for deciphering the function and the organisation of the genome, particularly if the strategies used have the goal of determining the order and time of gene expression, which represents another level of complexity in the analysis of biological organisms.

Proteomics (Protein Profiles)

A study of the proteome (proteomics, Table I) has been extensively used in the analysis of mammalian tissues and to a lesser extent in plants. The use of this technique has provided evidence of a particular expression pattern within a specific cell or tissue and has made a major contribution in human and microbial functional genomics. For example, differentially expressed proteins in melanoma cell lines and cancer cells have been described using 2-D protein gels; these protein profiles have been used to establish a tumour classification.

Table 1. Definitions used in genomics

Word	Concept
Genomics	The analysis of the genes contained in an organism
Proteomics	The analysis of protein
Proteome	The complement of proteins present in a cell
Transcriptomics	The analysis of transcript profiles
Transcriptome	The complement of transcripts (mRNA) in a cell

Following the extensive development in mammals and bacteria, there has been an increasing analysis of protein profiles in plants. The small number of studies include those on legume root nodules associated with nitrogen fixing bacteria; the lumenal and peripheral thylakoid membrane in pea; organelles of Arabidopsis;

the interconversion through a phosphorylation mechanism of two type of chloroplast RNA polymerase subunits; leaf development and defense responses in rice.

In addition, the protein pattern from xylem subject to abiotic stress has also been characterised, and this fingerprinting is being used to establish a QTL map from a maritime pine species. The presence of proteins not detected through transcript analysis, certain disadvantages are limiting the use of proteomics. These include the need for a more sensitive analytical system and the absence of an effective method for large-scale data comparison. Despite these limitations, there have been significant advances in the application of proteome analysis to genome analysis, particularly in describing the expression products in human and bacteria.

Transcriptomics (Transcript Profiles)

The analysis of transcripts expressed in different cell (transcriptomics) and tissues has been enormously facilitated by the development of Polymerase Chain Reaction (PCR). Several different strategies have been established to reveal expressed sequence tags (ESTs); their principles and applications will be discussed below. In general, the identification of a DNA encoded product is based on the synthesis of a copy DNA followed by PCR amplification.

Differential Display

Differential Display technique is based on the synthesis of cDNA, using an oligo dT primer (3') and an arbitrary oligonucleotide primer for the 5'-end. This specific condition makes it difficult to conduct analysis using the available databases. In an alternative protocol, arbitrary primers are used both for cDNA synthesis and PCR amplification. Both variants of RNA fingerprinting require a low annealing temperature during PCR amplification in order to achieve visualisable products. Consequently, the quantity of individual amplification products is not only a function of the initial concentration of that cDNA, but also is dependent upon the quality of a particular match between primer and template.

Abundant cDNAs with poor matches to the primers used are likely to outperform rare species with perfect matches during the course of PCR amplification. Despite this disadvantage, there are several reports of the description of genes expressed differentially in plants. These include the cloning of: the last enzyme (1 aminocyclopropane-1-carboxylate oxidase) involved in the synthesis of ethylene, a hormone that plays a role in fruit ripening and senescence in tomato; specific genes involved in the synthesis of flavonoids; genes differentially expressed in carpel development or in floral transition; modulators such as nitrate transporters concerned with plant/environment interaction; genes regulated by the light photoreceptor, phytochrome; and proteins involved in the response to abiotic stress. The technique has also been used to discriminate between individuals within a species.

cDNA-AFLP

This fingerprinting method relies on the selective amplification of a subset of DNA molecules from a more complex pool. The development of an amplified fragment length polymorphism (AFLP) provides a reliable tool to reveal the few differences between two close individuals. The system is based on the use of high stringency conditions, facilitated by adding double-stranded adaptors on the ends of restriction fragments, which serve as primer sites during amplification. Selective fragment amplification is achieved by adding one or more bases to the PCR primers which will only then be successfully extended if the complementary sequence is present in the fragment flanking the restriction site, thereby reducing the number of visualised bands.

The advantage of this strategy to reveal DNA fingerprinting has been presented in numerous reports—more than 260 in the last five years (PubMed database). Because of the advantage of this technique in generating a good individual polymorphism, cDNA has been used as template to amplify transcripts in plants. The use of any pair, of four and six base restriction enzymes, has been proposed as a method to generate differentially expressed products.

In fact, a combination of Pst I/Mse I or Ase I/Taq I has been successfully applied. This method seems to be more efficient as a way to discriminate and identify a particular RNA fingerprint; the cloning and subsequent characterisation of a specific band can easily be done, particularly if silver stain is used to develop the fingerprinting pattern.

Microarray

Recently a DNA chip technology has been developed to study gene expression profiles. Such techniques are based on the capacity to bind either DNA fragments or previously characterised oligonucleotides on a microscope slide. The plates are built by depositing specific DNA fragments at indexed positions using a computer controlled high speed robot with the specific DNA fragment sequences determined from any database originated including: cDNA clones, EST clones, anonymous genomic clones or DNA amplified from open reading frames (ORFs). Today's technology can display 409,000 spots in an area of 1.28 cm^2.

In theory, all 20,000-25,000 Arabidopsis genes could be displayed on a single slide. In fact, a slide with the Arabidopsis cDNA has been prepared for a large transcript analysis. The disposition of genes from the conifer Pinus taeda has also been prepared for use in physiological and molecular studies for wood formation. The technique is very sensitive, detecting mRNAs at level of 1/100,000 or 1/500,000. Despite the advantage of this precise technique, several problems have arisen and are expected to be resolved in the near future, including the high cost, identifying appropriate software for analysis of results, and standardising methods to allow comparison between results from different labs in light of the fact that in some systems the slide can be used only once.

Studies of genetic expression patterns have been done principally in humans and yeast. The determination of differentially expressed genes in cancer cells has been the primary goal of array efforts. Results from Saccharomyces cerevisiae include a large study on the modulation of differentially expressed genes involved in carbohydrate metabolism; 6,400 different genes were analysed simultaneously on a slide with the data providing valuable conclusions about the method of regulation of anaerobic respiration. Such technology applied to plants include those on Arabidopsis, soybean and maize, lima bean, and strawberry.

Serial Analysis of Gene Expression (SAGE)

Among the various techniques used to assess transcript abundance, the most powerful is probably SAGE. This technique, invented first to quantify gene expression in yeast, comprises the production of a short 10-14 nucleotide tag, with each tag representing a unique transcript present in a cell (all possible combination of 4 bases in a 10 bp sequence, 410 , gives more than 1 million different sequences). Determination of the sequence of a tag allows identification of the corresponding gene, and the frequency of a tag represents the steady state level of the mRNA from which it was derived.

The unique advantages of SAGE over alternative techniques for transcript analysis are: high sensitivity, threshold detection is one transcript in three cells; scalability, the technique can be used on any size of sample, from a few cells upwards; detection of all genes, including those of unknown function; avoidance of amplification bias; the data are digital and not derived (eg from an analogue fluorescent signal); the data are immortal and can be used at any time in a comparative study; data sets generated by one lab can be related directly to those produced by another.

These advantages have led to a rapid expansion of interest in using the method in a wide range of disease and developmental studies, almost all in humans. For example, in the last 18 months alone there have been approximately 50 published studies dealing with gene expression in cancers of the ovary, thyroid, skin, brain, bladder, esophagus, stomach, colon, breast and prostate, together with developmental studies on liver, kidney, muscle, skin, blood, nerve, bone, retina, and oocytes.

Such digital data sets are amenable to statistical analysis, and much of the data from these and other studies are now available in a SAGE site at the NCBI sponsored by the US National Cancer Institute. It is estimated that more than 200 medical research groups are now using SAGE methods for transcript profiling. The vast majority of plant scientists interested in large-scale gene expression studies seem to have only considered array methodology.

Applications to Transgenic Crop Improvement

All the information provided by the techniques described above can be used to investigate appropriate targets for modification by transgenic methods. The main objective of this strategy is to improve the quality traits of a particular crop intended to provide direct consumer benefit. The public resistance against using 'first generation' genetically modified organisms (GMO) has stimulated the development of GMOs that avoid the use of antibiotic markers.

The so-called 'second generation' GM products are currently in the research and development stage and may follow the first generation examples, which are primarily herbicides and insect-resistant products. Many such products are already undergoing field-testing in the USA, and there seems little doubt that in the future, many GM crops will include changes identified from transcriptome analysis. As an example of progress in this field, here focus on the new approaches to improve the resistance of crops to biotic and abiotic stress.

Stress-related Responses

There are many proposed transgenic routes for the improvement of stress-related responses in plants. Of the stress-related osmoprotectant compounds, the three most extensively studied are probably glycinebetaine, the sugar alcohol mannitol and trehalose, a sugar known to play a role in drought resistance of many organisms including the 'resurrection' plant. In bacteria this sugar is produced by the action of the two enzymes, trehalose phosphate synthase, which produces trehalose phosphate, and trehalose phosphatase, which degrades T-6-P into trehalose.

When one or both of these two enzymes are expressed in plants, the transgenics have larger leaves, altered stem growth, and improved response to stress. For example, when two selected transgenic tobacco plants grew under drought stress, their total dry weights were 28% and 39% higher than the controls. Chlorophyll fluorescence measurements showed a more efficient photosynthesis in the transgenic under these stress conditions. More recently, it has been claimed that similarly beneficial results can be achieved by modifying T-6-P via inhibition of endogenous trehalase—an enzyme that hydrolyses trehalose into two glucose moieties.

Transgenic tomato lines with various levels of trehalose are now being field tested. Additionally, over expression of the yeast HAL1 gene in tomato has been shown to have a positive effect on salt tolerance by maintaining a high internal K^+ concentration and decreasing intracellular Na^+ during salt stress, and production of fructans in transgenic plants has also improved their stress resistance.

Over expression of various glutamine synthases or glutamate dehydrogenase (GDH) is also claimed to improve growth and stress tolerance. Specifically, plants have been transformed with genes encoding the a- and β-subunits of the chloroplast-located GDH from the alga Chlorella sorokiniana. An alternative approach concerns the introduction of the uridine diphosphate glucose pyrophosphorylase gene from the bacterium Acetobacter xylinum. These transgenics, which have modified concentrations of cellulose precursors, are claimed to have increased growth rates and yield and improved response to stress conditions.

Similar improvements in performance are reported for tobacco plants overexpressing a cell wall peroxidase, and wheat or rice plants transformed with the barley late embryogenesis (LEA) gene. Modification of calcium related proteins is another method claimed to improve crop performance. For example, the introduction of a protein kinase domain-containing gene, a calcium dependent protein kinase gene, or a calcium/calmodulin-dependent gene, are all reported to be beneficial.

The most recent approach of this type is that involving the introduction of functional calcineurin activity as means of improving a salinity tolerant plant. Introducing a gene-encoding plant farnesyltransferase and enzyme inhibitors will enhance drought tolerance, improve resistance to senescence and modify growth habit when expressed in plants. Two other related strategies are based on the premise that many of the deleterious effects of stress are mediated through the accumulation of reactive oxygen species.

First, it has been suggested that the accumulation of these radicals is dependent on the presence of free iron in the cell, and consequently it has been proposed that controlling the amount of free iron could reduce oxidative damage. Convincing evidence of the value of this strategy wás shown in a study of transgenic tobacco expressing the alfalfa ferritin, an iron-binding protein. The second approach of this type is the recent claim that plants may be protected from the effects of stress-induced reactive oxygen species by the introduction of a transgene encoding an enzyme such as barley germin, an enzyme with oxalate oxidase and superoxide dismutase (SOD) activity that generates hydrogen peroxide.

This molecule is known to stimulate the plant's endogenous defenses, for example by the induction of PR proteins. Similar beneficial effects from modifying the levels of various SOD enzymes have been reported from other species. A specific heat shock protein has recently been implicated in thermal stress responses in Arabidopsis, whereas various proteins including choline oxidase and antifreeze proteins are claimed to be useful in protecting against low temperature stress. An additional recent method for providing general non-specific protection by up-regulating or pre-activating an existing defense pathway is exemplified by the introduction of a gene encoding the DREB1A transcription factor from Arabidopsis.

This factor is induced by a range of stresses and its introduction into transgenic Arabidopsis under the control of various promoters improved the tolerance to stress. The best results were achieved with transgenic expressing the gene under the control of the stress inducible promoter rd29A. Such transgenics showed better survival than the controls when exposed to salt, freezing and drought. The only field trial of material claiming to have salt tolerance is that conducted on Agrostis containing a betaine aldehyde dehydrogenase gene. Apart from the transgenic approach used in salt tolerance, the recognition of a particular marker can be used in direct selection on a specific progeny. Both strategies require basic information before selecting the way to make an improved crop.

8

Crop Biotechnology and Healthier Food Production

Food science and biotechnology can lead to substantial innovations in the production of healthier foodstuffs as well as increased profits by major food companies as in the period 2003-2004. The modification of vegetable oils is one of the key areas of plant and crop biotechnology, the overall objective being to increase their content in unsaturated fatty acids (mainly oleic acid) and to decrease that of saturated ones through conventional breeding, induced mutations or genetic engineering. Extensive work has been carried out on oilseed rape (canola), soybeans, peanut and sunflower with good results that led to the commercialisation of several products. Palm oil, which contains an equal proportion of saturated and unsaturated fatty acids, in addition to beta-carotene, is also a current research target, particularly of researchers at the Palm Oil Research Institute of Malaysia (PORIM).

In addition, replacing triglycerides with diglycerides in vegetable oils render them free of trans-hydrogenated fats and good cooking oils, e.g. 'econa oil' in Japan. Inulin and oligofructosans refer to a group of fructose containing carbohydrate polymers (fructans) which, in many plant species, act as protective agents against dehydration and cold temperatures and also offer many health benefits to humans, mainly in the stimulation of the growth of beneficial micro-organisms called bifidobacteria. These bacteria are sometimes used as a probiotic additive to foodstuffs such as yoghurt, as they can defeat harmful bacteria in the intestines and produce compounds with good health benefits.

These dietary fructans are also reported to have a lipid-lowering potential. They are not digested in the upper gastro-intestinal tract and therefore have a reduced caloric value. They share the properties of dietary fibres without causing a rise in serum glucose or stimulating insulin secretion. Inulin and oligofructosans can be used to fortify foods with fibre or improve the texture of low-fat foods without resulting in adverse organoleptic effects. Most of these two products currently on the market are either chemically synthesised or extracted from plant sources such as chicory roots. Oligofructosans are shorter chain polymers, highly soluble and provide 30 per cent to 50 per cent of the sweetness of sugar, and also have the other functional qualities of sugars. In formulation, inulin forms a smooth creamy texture, which makes this compound suitable as a fat substitute.

Oilseed rape (canola), which is a poor producer of inulin and oligofructosans, was used as model system. In particular, the production of two enzymes was to be evaluated: sucrose-1-fructose-1-transferase which adds a fructose moiety to a sucrose molecule, and fructan: fructan fructosyl transferase which continues to elongate the polymer by adding more fructose moieties to the chain.

Growers of nutraceutical plants need varieties with good agronomic potential and those that are consistent with the varieties in terms of germination time, height and maturity. Growers will need to be able to guarantee the quality of their natural health-beneficial products. Breeding methods can therefore be used to achieve uniform quality for clinical testing and for product development, as well as to remove these potentially harmful or otherwise undesirable compounds that are produced in the plants along with their therapeutic ones.

Long-chain polyunsaturated fatty acids are a research focus for nutritionists and food biotechnologists. Their beneficial effect on the functioning of the cardio-vascular system has been initially mentioned since the 1970s in the medical literature. Long-chain polyunsaturated fatty acids belong to two main categories: omega-3 (first double bound at carbon 3 on the chain) and omega-6 (first double bound at carbon 6 on the chain). Among omega-3 fatty acids, there are the alphalinolenic acid (ALA) with 18 carbon atoms, eicosapentaenoic acid (EPA) with 20 carbon atoms and docosahexaenoic acid (DHA) with 22 carbon atoms.

The human body cannot synthesise the ALA as well as the linoleic acid which is an omega-6 fatty acid. Omega-3 fatty acids are found in rapeseed and soybean oils (linolenic acid), marine animals and human milk (EPA, DHA). Food-consumption surveys carried out in France have shown that the consumption of omega-3 fatty acids was insufficient and the ratio of omega 6 to omega 3 was not balanced (this ratio should be between 5 and 10). Although research is being carried out on the precise role of these fatty acids on human health, it is not easy for the public to have a clear view of established scientific facts and amid controversial statements.

Let us look now at what maybe causing confusion among the public as regard the issue of omega-3 fatty acids. It may have begun with the study that revealed lower morbidity and mortality due to cardio-vascular of Greenland's Inuits who consume a lot of fatty fish. In France, the French Agency for Food Sanitary Safety (AFSSA) convened a meeting of experts on the effects of omega-3 fatty acids on the cardiovascular system. They concluded that the supplementation of daily diet with these fatty acids could have a beneficial impact on the functioning of the cardio-vascular system, as a secondary prevention measure.

Morbidity and mortality reduction was indeed significant among the persons who suffered form cardio-vascular or metabolic diseases. However, omega-3 fatty acids did not act on cholesterol; they may act on triglycerides and cell membranes, as well as on blood clotting and heart excitability; they may also have, through prostaglandins (some of these acids are precursors in the biosynthetic pathways of prostaglandins), a positive effect on hypertension. The experts also warned against the role of the consumption of excessive quantities of omega-3 fatty acids, as they would increase cell susceptibility to free radicals. They recommended a maximum daily intake of EPA and DHA of 2g per day.

Then there are also the claims on the prohibitive effects of omega-3 fatty acids on tumours. To this, the experts concluded that all the studies carried out up to 2004 on food habits did not substantiate in humans any evidence indicating that an enrichment of the diet with precursors of omega-3 fatty acids would protect against cancer. However, research work carried out on rats has shown that a diet enriched with omega-3 fatty acids caused a 60 per cent decrease in size of mammary tumours, twelve days after radiotherapy, compared with a 31 per cent decrease in animals fed with a non-enriched diet. Trials are expected to be carried out on humans.

Given the insufficiency of evidence, the benefits of taking Omega-3 pills remain inconclusive. In view of this, the general advice is to consume fish at least twice a week. The same goes for rapeseed oil. This is sufficient to meet the daily needs of omega-3 fatty acids. It is also recommended to feed poultry with rapeseed meal rather than with sunflower meal, because the former is richer in omega-6 fatty acids. Thus, consuming this kind of poultry meat would provide enough omega-6 fatty acids.

Food Crops Biofortification

Biofortification of food crops makes sense as part of an integrated food-systems approach to reducing malnutrition. It addresses the root causes of micronutrient deficiencies, targets the poorest people, and is scientifically feasible and cost-effective. It is a first step in enabling rural households to improve family nutrition and health in a sustainable way. HarvestPlus is a coalition of CGIAR Future Harvest Centers or Institutes, partner collaborating institutions and supportive donors (The Bill and Melinda Gates Foundation, DANIDA, Swedish International Development Assistance—SIDA, US Agency for International Development—USAID, and the World Bank).

The International Center for Tropical Agriculture (CIAT) and IFPRI are coordinating the plant breeding, human nutrition, crop dissemination, policy analysis and impact activities to be carried out at international Future Harvest Centers, national agricultural research and extension institutions, and departments of plant science and human nutrition at universities in both developing and developed countries.

An initiative of the Consultative Group on International Agricultural Research (CGIAR), HarvestPlus is a global alliance of research institutions and implementing agencies coming together to breed and disseminate crops with improved nutritive value (biofortification), e.g. with a higher content of iron, zinc and vitamin A. The biofortification approach is backed by sound science. Research on this funded by the Danish International Development Assistance (DANIDA) and coordinated by the International Food Policy Research Institute (IFPRI) led to the following conclusions:

- substantial, useful genetic variation exists in key staple crops;
- breeding programmes can readily manage nutritional quality traits, which for some crops have proven to be highly suitable and simple to screen for;
- desired traits are sufficiently stable across a wide range of growing environments; and
- traits for high nutrition content can be combined with superior agronomic traits and high yields.

Initial biofortification efforts as of 2005 will focus on six staple crops for which prebreeding studies have been completed: beans, cassava, sweet potatoes, rice, maize and wheat. The potential for nutrient enhancement will also be studied in ten additional crops that are important components in the diets of those with micronutrient deficiencies: bananas/plantains, barley, cowpeas, groundnuts, lentils, millet, pigeon-peas, potatoes, sorghum and yams.

During the first four years (1 to 4) of the project, the objectives are to: determine nutritionally optimal breeding objectives; screen CGIAR germplasm for high iron, zinc and beta-carotene amounts; initiate crosses of high-yielding adapted germplasm for selected crops; document cultural and food-processing practices, and determine their impact on micronutrient content and bioavailability; identify the genetic markers available to facilitate the transfer of traits through conventional and novel breeding strategies; carry out in-vitro and animal studies to determine the bioavailability of the enhanced micronutrients in promising lines; and initiate bio-efficacy studies to determine the effect on biofortified crops on the micronutrient status of humans.

During the following three years (5 to 7), the objectives are to: continue bio-efficacy studies; initiate farmer participatory breeding; adapt high-yielding, conventionally bred, micronutrient-dense lines to select regions; release new conventionally-bred biofortified varieties to farmers; identify gene systems with potential for increasing nutritional value beyond conventional breeding methods; produce transgenic lines at experimental level and screen for micronutrients, test for compliance with biosafety regulations; develop and implement a marketing strategy to promote the improved varieties; and begin production and distribution.

During the last three years of the project (8 to 10), production and distribution of the improved varieties will be scaled up; the nutritional effectiveness of the programme will be determined; and the factors affecting the adoption of biofortified crops, the health effects on individuals and the impact on household resources will be identified.

Staple Crops and their Biotech Status

Rice is the dominant cereal crop in many developing countries and is the staple food for more than half of the world's population. In several Asian countries, rice provides 50 per cent to 80 per cent of the calorie intake of the poor. In South and South-East Asian countries, more than half of all women and children are anaemic; increasing rice nutritive value can therefore have significant positive health impact.

Food-consumption studies suggested that doubling the iron content in rice could increase the iron intake of the poor by 50 per cent; germplasm screening indicated that a doubling of iron and zinc content in unmilled rice was feasible. Milling losses vary widely by rice variety, with losses of iron being higher than losses of zinc, which suggests than more zinc is deposited in the inner parts of the rice endosperm. Under the HarvestPlus project, improved rice germplasm will be provided to national partners in Bangladesh, Indonesia, Vietnam, India and the Philippines.

The improved features will be incorporated into well-adapted and agronomically-preferred germplasm in ongoing breeding programmes at the national and regional level. A plant-biotechnology approach is the current priority for enhancing provitamin-A content of the rice endosperm. The leading varieties will be field tested for agronomic performance and compositional stability in at least four countries.

Wheat: The International Maize and Wheat Improvement Center (CIMMYT) is leading the HarvestPlus research endeavour on wheat biofortification in order to increase people's intake of iron and zinc. Given that spring wheat varieties developed by CIMMYT and its partners are used in 80 per cent of the global spring wheat area, the potential impact of iron-enhanced wheat could be dramatic. The initial target countries will be Pakistan and India, in the area around the Indo-Gangetic plains, a region with high population densities and high micronutrient malnutrition.

The highest contents of iron and zinc in wheat grains are found in landraces of wild relatives of wheat such as Triticum dicoccon and Aegilops tauschii. Because these wild relatives of wheat cannot be crossed directly with modern wheat, researchers facilitated the cross between a highmicronutrient wild relative, Aegilops tauschii, and a highmicronutrient primitive wheat, Triticum dicoccon, to develop a variety of hexaploid wheat that can be crossed directly with current modern varieties of wheat and have 40 per cent to 50 per cent higher contents of iron and zinc in the grain than modern wheat.

The first biofortified lines will be delivered to the target region by 2005, i.e. broadly-adapted, high-yielding, disease-resistant wheat lines. The first highyielding lines with confirmed iron and zinc contents in the grain should be available for regional deployment by mid-2007.

Maize: Maize is the preferred staple food of more than 1.2 billion consumers in sub-Saharan Africa and Latin America. Over 50 million people in these regions were vitamin A-deficient in 2004. The International Maize and Wheat Improvement Center (CIMMYT) and the International Institute of Tropical Agriculture (IITA) are identifying micronutrient-rich maize varieties and will carry out adaptive breeding for local conditions in partnership with National Agricultural Research Systems (NARS) in Africa and Latin America.

The project under HarvestPlus is initially focusing on maize varieties having increased contents of provitamin A because a useful range of genetic variation has already been identified for this trait. The first target countries are Brazil, Guatemala, Ethiopia, Ghana and Zambia. To support the breeding programme, research is being conducted in Brazil, the USA and Europe to develop simple, inexpensive and rapid screening protocols for provitamin A, so as to reduce the cost of assays from $70-100 to $5-10 per sample.

Research in Brazil and the USA is also focused on finding genetic markers to facilitate marker-assisted selection for provitamin A concentration. In collaboration with the University of Wageningen, a human efficacy trial was planned with provitamin A-rich maize in Nigeria for 2005 in order to study provitamin A retention or loss for different storage, processing and common cooking methods. To facilitate extension and dissemination of biofortified maize varieties, country teams will be formed in the target countries in order to conduct adaptive breeding research, farmer-participatory variety evaluations, nutritional advocacy and promotional activities.

Beans: Common beans (Phaseolus vulgaris) are the world's most important food legume, far more so than chickpeas, faba beans, lentils and cowpeas. For more than 300 million people, an inexpensive bowl of beans is the main meal of their daily diet. The focus of HarvestPlus research is on increasing the concentration of iron and zinc in agronomically superior varieties. Over 2,000 accessions from the International Center for Tropical Agriculture (CIAT,) gene-bank and several hundred collections of African landraces have been screened for their nutrient contents.

While the average iron concentration in these varieties is about 55 mg per kg, researchers have found varieties the content of which exceeds 100 mg per kg. The eventual goals are to obtain favourable combinations for productivity and nutritional traits, double the iron concentration and increase zinc concentration by about 40 per cent. The first bred lines with 70 per cent higher iron will likely emerge in 2006, while lines with double concentration of iron are anticipated in 2008.

The proportion of iron and zinc that can be absorbed from legumes such as common beans is typically low, due to anti-nutrients, specifically phytates and polyphenols, which normally bind to the iron and zinc, making them unavailable to the organism. Research indicated that it might be possible to reduce polyphenol concentrations genetically, thereby improving iron bioavailability. In contrast, vitamin C is an iron-absorption enhancer because it binds to iron and prevents it from becoming attached to the iron-absorption inhibitors. Beans are often consumed with vegetables, including bean leaves with the potential of bean leaves as a source of vitamin C still to be explored.

Cassava: Cassava, also known as manioc or tapioca, is a perennial crop native of tropical America that is also widely consumed in sub-Saharan Africa and parts of Asia. With its productivity on marginal soils, ability to withstand disease, drought and pests, flexible harvest dates, cassava is a remarkably adapted crop consumed by people in areas where drought, poverty and malnutrition are often prevalent.

Cassava is typically white in colour and, depending on the amounts of cyanogenic compounds, can be sweet or bitter. The International Center for Tropical Agriculture (CIAT) will coordinate HarvestPlus' overall activities on cassava biofortification and be primarily responsible for research in Asia, Latin America and the Caribbean. The IITA will be responsible for cassava biofortification in Africa. In collaboration with the University of Campinas, São Paulo State, Brazil, the total content of provitamin A in cassava varieties (roots) will be determined spectrophotometrically.

Provitamin-A retention studies will also be carried out on different preparation and cooking methods used in cassava consuming countries. A method for storing cassava roots for several weeks or a few months is needed for programmes quantifying hundreds of samples per year. Initial data suggest that the anti-oxidant property of a few yellow pigments in cassava roots may delay physical deterioration of the roots. The longer shelf life of yellow cassava roots may not only appeal to farmers and consumers, but may also increase the demand for biofortified varieties.

Nutritionally improved germplasm coupled with superior agronomic performance can be developed as a medium term apprcach with products reaching the farmers as soon as 2009. The aim is to identify and select, from the varieties having both high provitamin-A contents and good agronomic performance, those with the highest iron and/or zinc content.

Sweet potato: Sweet potato is an important part of the diet in East and Central Africa where vitamin-A deficiency is widespread. At present, African predominant sweet potato cultivars are white or yellow-fleshed varieties that contain small amounts of provitamin A. In contrast, the orange-fleshed varieties are believed to be one of the least expensive, rich, year-round sources of provitamin A. Boiled orange-Fleshed sweet potato, such as the Resisto variety developed in South Africa, contains between 1,170 and 1,620 Retinol Activity Equivalents (RAE) per 100 g and is estimated to provide between 25 per cent and 35 per cent of the recommended daily allowance for a preschool child.

Experts at the International Potato Center (CIP), who developed a biofortified orange-fleshed sweet potato, estimated that when fully disseminated, this sweet potato could reduce vitamin-A deficiency in as many as 50 million children. To encourage a switch from non-orange to orange-Fleshed varieties, the texture of the latter must be changed because they tend to have a high-moisture content and adults prefer varieties with a low water content, i.e. a high dry biomass. Plant breeding is ongoing to increase the dry biomass of the provitamin A-rich orange varieties, to improve organoleptic characteristics and at the same time improve their resistance to viruses and drought.

About 40 varieties of sweet potato with high dry biomass and provitamin-A content have been introduced to sub-Saharan Africa. Of these, 10 to 15 were being tested widely in different agro-ecological areas in some countries. Some original varieties, mainly local landraces, have been well accepted by farmers and were being distributed on a small scale.

Issues in Healthy Foodstuffs

Innovation in healthy foodstuffs is also fraught with costly failures. For instance, Procter & Gamble spent 30 days developing Olestra, a fat which the digestive system cannot absorb. But the product has been dogged by claims that it inhibits the absorption of vitamins and nutrients that may help prevent cancer; by a hostile lobby group, the Centre for Science in the Public Interest; and by regulatory problems.

In 1996, after eight years of tests, the US Food and Drug Administration (FDA) allowed it to be used as an ingredient, but products made with it had to carry the warning that it might cause gastro-intestinal distress. In the summer of 2003, the FDA allowed the warning to be taken off advertisements. Moves toward healthier products and functional foodstuffs also fuel the professional lives of lawyers, regulators and stock-market analysts.

Many companies have recently appointed advisory boards composed of top nutritionists. The FDA itself is acquiescing to companies' proposals to include in their products' health benefits in their labels. The FDA has, in fact, liberalised the rules on making health claims, adopting a four-tier system enabling consumers to decide based on how solid the science is behind any particular product's health claim. Calcium ability to protect against osteoporosis, for instance, is reckoned very solid while omega-3 fatty acids to prevent heart diseases are considered good, but second level.

By early 2003, the FDA announced that from 2006 consumers must be informed of the amounts of trans-fats in foods notwithstanding the already wide publicity on the adverse effects of transfats. Health authorities in Europe are also striving to regulate this kind of research and to establish marketing standards. All the difficulties relate to the need to demonstrate the impact of these foodstuffs on disease prevention in humans.

Genetic engineering is useful for producing crops or food ingredients deprived of some undesirable elements or enriched with healthy substances, and therefore qualified as nutraceutics. To be attractive to the consumers, these foodstuffs should not be too expensive. Ageing populations are a particular target for nutraceutics, which can play a key role in the nutrition of old people suffering from under- and malnutrition. Between 1998 and 2002, it was estimated that the annual turnover of modified milks increased by 10 per cent in Europe and 36 per cent in the USA among people of more than 65 years.

Agri-food companies are also designing communication policies not just for consumers but also for physicians, pediatricians and nutritionists, like the pharmaceutical groups, in order to highlight the benefits of their products. These policies have to take into account the cultural differences with regard to food and nutrition among the countries. They should also state the preventive role of nutraceutics as well as their therapeutic effects.

Genetic Markers with Beverage and Food Quality

Correlation of genetic markers with meat quality: In 2002, the Maryland (Savage)-based biotechnology company, MetaMorphix, acquired the livestock genotyping business of Celera Genomics, a company founded in 1998 to sequence the human genome; it then joined up with Cargill, Inc., to commercialise a genetic test that will help to reveal, prior to slaughter, a cow's propensity to produce desirable meat. That task is being carried out by analysing thousands of so-called single-nucleotide polymorphisms (SNPs) in the bovine genome.

A SNP is a place where the genomes of individual animals vary by a single nucleotide. SNPs are therefore convenient marker versions of particular genes, and different versions of genes result in differences between animals. MetaMorphix and Cargill, Inc, tried to find out which SNPs were associated with variations in meat quality, such as flesh colour, amount of marbling, wetness and tenderness, so that these could be identified before slaughtering an animal, and suitable animals will thus be reserved for breeding.

In 2002-2003, Cargill, Inc., studied 4,000 cattle, trying to correlate MetaMorphix's genetic markers with meat quality—and with other important traits, such as growth rate. Almost 100 useful SNPs have been identified from this study. As a result, a prototype testing kit was to be used by the firm as of August

2004. The first 'designer meat' produced this way was expected to be marketed in 2005.

DNA Fingerprinting of Grapevine Varieties

Since 1990, US and French researchers have been trying to establish the phylogenetic tree of the varieties of grapevine grown throughout the world, using the fingerprinting technique. The research consists of analysing the structure of certain regions of the genome of some grapevine varieties and comparing it with that of other varieties, so as to establish possible phylogenetic relations.

For this analysis, the DNA is extracted from young ground leaves, but also from fruits and branches. The results of this first research have been published in 1999 in the Science journal. There are about 6,000 grapevine varieties cultivated worldwide. In order to establish the origin of a variety, specialists used to rely on phenotypic traits, such as the morphology of leaves, berries and grapes. In this way, varieties could be grouped in a few families.

DNA fingerprinting enables the researchers to go further. It appears that the current grapevine varieties are the remote offspring of the grapevines grown during the Antiquity around the Mediterranean, or during the European Middle Ages. The current varieties are the result of lengthy breeding activities, identification and comparison work, and stabilisation of the selected strains or lines.

The cooperation between the Californian team and the French specialists of the National Higher School of Agronomy, Montpellier, associated with the Genetic Research and Breeding Unit of the National Agricultural Research Institute (INRA), has led to undisputable results concerning the origins of various grapevine varieties: the chardonnay (the most famous and expressive variety in Burgundy), aligote (also from Burgundy), gamay (red variety) and melon of Burgundy (which has colonised the area of muscadet) are all cousins and almost siblings.

Such a conclusion, derived from the analysis of DNA fingerprinting, was not a surprise for oenologists and tasters because the wines made from these four varieties share common structures and aromas. This kinship is particularly expressed in the ageing wines. But more surprising than the discovery of that kinship among the four grapevine varieties, was the identity of one of the progenitors of the initial couple that gave rise to these varieties. Indeed, several historical elements were in favour of the creation of lines through the cross pollination between the pinot noir and the white gouais; these crosses have given birth—as proven by the fingerprinting analysis—to the three white and red varieties grown for a long time in various French provinces, and for some decades, in many regions of the globe.

The surprising aspect of this discovery was that the white gouais is almost unknown, although the vine specialists in Montpellier continue to grow it and to make wine form it for their own pleasure.

The white gouais was also formerly grown in the Jura and Franche-Comté. For the US and French researchers, this variety which has played a key role in the history of vine and wine, is the same as the heunisch variety of Central Europe, introduced in Gaul by a Roman emperor originating from Dalmatia. In Montpellier, the French researchers participating in the joint study with the US scientists, Davis, have tried to reproduce the breeding between the pinot noir and white gouais in order to seek confirmation of the genetic research.

Other attempts were expected to widen even more the range of cultivated grapevine varieties, for both their fruits and wines derived from them. But this approach was hindered by a drastic regulation, which practically prohibits any venture of this kind, while non-French winemakers and vinegrowers could do it. In Apulia, in the heel of the Italian boot, drawing on grapes grown by up to 1,600 small farmers in the area, a California wine consultant associated with another Italian wine consultant from Friuli are producing and marketing wines that have scored a great success worldwide, with 2004 projected sales 15 times as big as those in 1998, the winery's first year. The wines are called A-Mano—handmade—and by far the best known is a robust red made from a once-obscure grape named primitivo.

DNA testing by Carole Meredith at the University of California, Davis, established that primitivo is a descendant of a grape called crljenak kastelanski, widely known in the 18th and 19th centuries on the Dalmatian coast of Croatia (a crljenak cross with dobrinic, plavac, mali, is being grown on that area today). California's zinfandel, she showed, is genetically the same as primitivo, though how it crossed the ocean remains a subject of dispute.

Apulian primitivo and zin are not twins, of course; climate, soil and vinification all help to shape a wine's look, aroma and flavour, along with the grape variety. But the two share several characteristics: both are fruit-rich, chewy, sometimes lush wines, a deep violet-red in colour, often too high in alcoholic content for comfort, but much more subtle if carefully handled.

Organic or Biological Agriculture

The phrase "organic or biological agriculture" designates an agricultural mode of production that does not rely on the use of chemicals, e.g. fertilisers and chemical pesticides. It also excludes any genetically modified organism, and is labour intensive. It does not mean that the products are necessarily of a higher quality than those derived from conventional agriculture. How much acreage is dedicated to this version of farming?

It depends on the definition, but according to an extensive survey by the Germany-based Ecology and Agriculture Foundation, in 2003 Australia led with 7.6 million hectares, followed by Argentina (3 million ha) and Italy (1 million ha). When expressed as a percentage of total arable land, these figures are far from stunning. The highest proportions are found in Europe where organic agriculture methods are well defined and products registered.

Thus Austria, the leading European country in biological agriculture and products, devotes 11.3 per cent of its total arable land to this kind of agriculture. In France, the proportion is only 1.4 per cent and 11,000 farmers are considered organic producers. It is Europe's first-biggest agricultural country occupying 13th place with regard to biological agriculture. The methods of biological agriculture have been applied in France since the 1950s, but it was only at the end of the 1970s that small groups of farmers began to organise themselves, e.g. to create the National Federation of Biological Agriculture (FNAB).

In the United Kingdom, demand for organic food is growing at over 40 per cent a year. This demand amounted to nearly $1.6 billion in 1999, with a third of the population buying organic food. In France in 2002, 65 per cent of French people had consumed a 'bio' product during the last 12 months, compared with 50 per cent in 2001 and 40 per cent in 2000. In 2003, however, the French market of organic foodstuffs and products, labelled as "bio", after growing at a 20 per cent annual rate, showed signs of slowing down.

Although those responsible for producing and transforming organic products tended to underestimate this stagnation, the growth rate of the 'bio' products market was estimated at 6 per cent to 10 per cent in 2003. According to the institute TNS Media Intelligence which scrutinises the evolution of consumers' behaviour, 57 per cent of French housekeepers had bought at last one 'bio' product in 2003 a low figure compared with the levels of consumption in the United Kingdom (80 per cent), Italy and Germany.

There are several reasons for this stagnation. In the case of milk, the price of the litre has fallen by 0.02 to 0.08 cents since early 2002. Meat profits were 20 per cent to 30 per cent less. Producers of poultry, vegetables and fruits suffered from crises comparable to those affecting their colleagues in conventional agriculture. However, there did not seem to be an overproduction, because although consumption of 'bio' products was decreasing, it was not met by French production.

In 2002, ironically, 21 million litres of milk produced in France were 'declassified' and sold through conventional outlets, while 'bio' products sold in France were imported, e.g. fruits, vegetables, cereals and exotic products. Most of stakeholders involved in organic farming and production, as well as analysts, agree that the crisis is rather a marketing issue; in other words, 'bio'-product producers did not have the full capacity to put these products in the conditions when the consumers want to find them. The French group Biolait that collects onefourth of 'bio' milk produced in France failed to impose its prices to the transformers who went away; henceforth the declassification of this milk into conventional one, and the company Lactalis whose supply broke down, had to import its 'bio' milk from Germany. Meanwhile, there are also problems of distribution and price, as well as certification and fraud issues.

Although hyper and supermarkets are the first selling places of 'bio' products (52 per cent of purchases are made in these places), the French Carrefour supermarkets sell about 300 products, while the British Sainsbury plc offer a much wider range of 1,200 products. Specialised shops are also less than 1,000 in France. Consequently customers have some difficulties finding 'bio' products in their usual shopping places. However, partnerships have been established between these retailers and the farmers.

Some farmers are also trying to diversify their outlets so as not to depend on the supermarkets only: thus, in France BioBourgogneViandes, created in 1994 near Dijon and comprising some forty cattle raisers, is selling meat to individual clients directly (for an 18-kg order, the client pays •12 per kilo and the transport cost). BioBourgogneViandes has also purchased a first butcher's shop in the central market of Dijon, while the number of members grew to 70 by the end of 1999, with an annual turnover of •2.4 million. The cattle raisers owned four shops in different villages and were selling, in addition to meat (by correspondence), such biological products as wine, cheese and vegetables.

Pricing: The price of 'bio' products remains the principal obstacle to their purchase. According to their survey carried out in October 2003 by the review 60 millions de-consommateurs in France, 'bio' products sold in hyper- and supermarkets cost 40 per cent to 60 per cent more than conventional products, and 70 per cent to 100 per cent more in open air markets and specialised shops. It is true that organic farming needs more labour, particularly for growing vegetables and fruits. Also the small volumes of milk and meat make the harvest, bottling and transportation of these products very costly, because one has to find the industrial tools (rather few) that meet the stringent standards of this form of production.

Only 'bio' eggs, laid by hens raised in open backyards and fed with foodstuffs derived from organic farming, have met with great commercial success. There is also the issue of economies of scale. The high costs of 'bio' products and the need to successfully compete with conventional products have led organic farmers in Europe to request assertive policies from the respective governments in favour of organic agriculture, starting with a marked increase in the acreage devoted to it.

Germany, Italy, Denmark and Austria have designed public policies to support the growth of organic farming, while in France it has been suggested that the acreage devoted to organic farming should reach 3 per cent of total arable land, in order to become economically sound.

Certification: Organic farmers defend an alternative agricultural, economic and social model. They prohibit the use of genetically modified organisms and demand, in case the ban on transgenic crops is actually lifted in Europe, a threshold of 0.1 per cent for an adventitious presence of GMOs in their crops, instead of 0.9 per cent as it has been decided for conventional crops. The requirements for organic farming are applicable throughout Europe, with some national differences that may create distortions in competition.

In France, the requirements for livestock husbandry are the most stringent, as all the feed should be produced on the farm itself. The AB logo, which applies to 'bio' products in France, can be applied to non-French organic products provided that they meet the national requirements. A common European logo also exists to label those 'bio' products. These logos do certify the mode of production of the product, but not its quality as does the Red Label in France.

Regarding exotic products, they can also be labelled with the same AB logo, but they raise problems of traceability and control. Many developing countries are devoting an increasing acreage of their arable land to organic farming, as they are attracted by premium prices on the international market. This is, for instance, the case of Chile, which has important outlets for its organic products in the European Union and Japan. Surinam and Papua New Guinea head the list of African, Caribbean and Pacific (ACP) countries as 'bio' producers. Demand in the Western countries for organic fruits and vegetables is enticing producers throughout the ACP countries to be organised and establish their foothold in the market of opportunity.

In early October 2001, more than 170 traders, producers, researchers and support agencies converged on Port-of-Spain, Trinidad and Tobago, to do that at a conference on diversifying regional exports through developing organic agriculture. They came from Cameroon, Malaysia, 17 Caribbean nations and departments, 11 countries of Central and South America, and eight countries of North America and Europe. The conference launched a new study on World markets for organic fruit and vegetables by the FAO, the International Trade Centre of the United Nations Conference on Trade and Development (UNCTAD) and World Trade Organisation. Debates at the conference led to concrete proposals for national standards, regional certification, information services and special measures for smallholders wanting to switch to organic farming.

The certification of organic or 'bio' products is part of a wider trend that consists of drawing the consumers' attention to the origin and quality of their foodstuffs and beverages. This trend responds to the concern that quality is threatened by industrialisation of food production and processing. Although reaching a high quality standard may require years of work and great financial endeavours, farmers are interested in following suit. In France, for instance, one-third of farmers, 6,700 companies and 6,000 distributors were, in 2001, engaged in official procedures aimed at certifying their products. Certification in France deals with the quality and origin of the product.

European certification includes a certified protected origin (AOP), the French AOC, as well as the protected geographic indication, which establishes a less strict geographical relationship than the AOP. In the United Kingdom, a leading company in the production of 'bio' products is Duchy Originals, founded by the Prince of Wales and established at the Home Farm of Tetbury, near the Prince's residence of Highgrove (Gloucestershire). It sells chocolates, bread, honey, biscuits, cheese, ham, sausages and soft drinks. On 23 March 2004, a new product has been launched: a 'bio' shampoo.

The latter, bearing the logo Houmont, contains rose and lemon essences and will be a strong competitor of Body Shop's comparable products. It is a joint venture between the Prince of Wales and a famous London-based hairdresser. Duchy Originals is a prosperous enterprise, with a •22-million turnover in 2003-2004 and profit reaching •343,000. The 'Prince Charles' label is found in big department stores as well as worldwide, particularly in the USA and Commonwealth countries. Benefits are transferred to a charity foundation chaired by the Prince of Wales. Since its creation in 1990, about £2 million had been transferred to philanthropic associations by Duchy Originals which employed 100 people.

Official certifications are facing the competition of private certifications, such as the indication of local or regional brands, of provenance, emphatic mentions, which do not necessarily represent a qualitative content or value and which are not submitted to an independent review. Monitoring and control of 'bio' products are carried out in France by some thirty certifying bodies, except the AOCs which are under the control of the National Institute of Denomination of Origin. All these bodies are authorised by the state and their action is followed by that of the General Directorate for Competition, Consumption and Fraud Repression.

The denomination of origin label (AOC) is not peculiar to rich countries. Thus, China, with the assistance of France, has adopted a law on appellations in 2000 and has created about 30 AOC labels concerning yellow wine, teas and hams that are typical for some regions. Vietnam has followed suit in 2001; the first geographic indication regarding the 'nuoc mam of Phu Quoc' has attracted the interest of Unilever which invested $1 million for transferring its production in that island; since then, the price of this AOC nuoc mam has trebled. Sixteen countries of West Africa have requested the French INAO to identify two products in each of them that could be certified.

Morocco wants to protect its argan oil (extracted from the seed of the tree Argania spinosa, an endemic species of the south-west of Morocco), while Bolivia wishes to label its wine and quiñoa—a nutritious seed from Amaranthus quiñoa—and Brazil wants to tag its best wines. These efforts demonstrate that the globalisation of nutrition and food should not necessarily lead to homogenisation of products but to the promotion of trade relations that respect nutritional differences and cultures.

Segregation: In the fall of 1999, two French industrial corporations, Glon-Sanders—the leader in animal feed and egg production in France—and Bourgoin—the European leader in poultry production—decided to set up a non-genetically-modified (GM) soybean chain, through which this legume was traced during the whole transformation process, from the seeds to the eggs and poultry sold to retailers. Eleven cooperatives, including 2,000 farmers and representing 12,000 hectares, responded positively to both industrialists in order to produce the so-called 'soja du pays'.

In 1999, the expected harvest was 50,000 tons of beans (i.e. 20 per cent of French production) and in 2000 more than 100,000 tons. The soybean meal (i.e. crushed soybeans the oil of which is extracted) or extruded beans (i.e. soybeans processed to make them digestible) would be used to feed poultry, sold under the Duc label as well as laying hens. It involved two production zones were concerned: one in the south-west of France, including eight cooperatives and the company, Céréol, which processes soybeans into meal; the second in Burgundy, including three cooperatives and the company Extrusel, specialised in the production of extruded seeds.

In order to mitigate the risk of contamination of locallyproduced soybeans by imported US beans, a cooperative from Castelnaudary (southwest of France) the Groupe Occitan, checked the French origin of seeds, isolated the production plots, stored the harvested beans separately until they were delivered. Regarding Extrusel, a subsidiary of four grain cooperatives of Burgundy and Franche-Comté and of two livestock-feed producers of Saône-et-Loire, it produced 20,000 tons of extruded soybeans per annum (annual turnover of about $6.2 million). Extruded soybeans are a very digestible feed which supplies both proteins and fats and is incorporated into poultry and hog rations.

In February 1999, Extrusel made the decision to only use soybeans of which the non-GM status could be guaranteed. During the spring of 1999, 8,000 hectares were sown with soybeans that were certified as non-genetically-modified by about one thousand farmers; this acreage represented the whole cultivated area between the cities of Belfort and Mâcon. A fully operational traceability system was set up from the farm to the client. Part of the beans was produced in the company's station and the rest was bought from outside suppliers with guaranteed origin, so as to be in conformity with Extrusel ISO 9002 standards.

Farmers who subscribed to this new productive venture and who harvested their first non-GM soybeans in the 1999 fall, sold their product at 1,100 Francs per ton, compared with 1,000

Francs per ton of 'ordinary' soybeans and 1,300 Francs per ton of beans used for human consumption. Bourgoin imposed additional constraints on the farmers that produced the 'soja du pays'. They must not use genetically modified soybeans and they should trace the production of the beans at all stages of the process, they must grow them at a certain distance of pollution sites (e.g. chemical factories and incineration centres which could generate dioxin residues) and should refrain from spreading sludge originating from wastewater treatment.

Certified poultry represented almost 85 per cent of annual total turnover of Bourgoin-Duc, and through the company's decision not to use transgenic soybeans it wished to anticipate consumers' demand. Bourgoin-Duc had already prohibited the use of transgenic maize since 1996 in the feed used for poultry. In addition Bourgoin had tried, with food distributors such as Carrefour, to set up non-GM soybean production chains in the USA and Brazil.

Fraud: Biological agriculture is not free from criticism because of fraud. By early 2000, in France, the agriculture ministry's Directorate-General for Competition, Consumption and Fraud Repression carried out an enquiry on false biological cereals. About a dozen important operators were involved in the following traffic: a dealer buys conventional cereals and establishes forged certificates that qualify them as derived from biological agriculture; the cereals are sold as 'bio' products either to feed producers or directly to livestock raisers.

As the selling price of 'bio' cereals could be twofold of that of conventional cereals, the illegal profit could vary from 1 Franc to 50 centimes per kilo, and because of the volumes concerned the benefits could be very high. The French authorities discovered an international network involving in particular Italian capital. Conventional cereals were sometimes purchased in France and shipped—really or virtually—to Italy, Belgium or the Netherlands, from where they returned with the 'bio' label.

Another traffic was initiated in Central Europe, particularly in Romania or Ukraine, and the cereals were transferred to

France. On 23 March 2000, the French inspectors spotted a society based in Brittany (Carhaix), Eurograin, which they suspected of having marketed 50,000 tons of cereals of doubtful origin in 1998 and 1999. Earlier, on 3 March 2000, a public enquiry had been opened in the Vienne Department regarding 12,000 tons of cereals commercialised by Bio Alliance, a company based in Chasseneuil-du-Poitou. Its manager was condemned for having unduly used a 'bio' label on bovine meat.

Another enquiry concerned the shipment of the Celtic Ambassador, a boat inspected in 1997 in Bordeaux; the 4,500 tons of cereals found in the boat officially originated from Romania, and had been certified 'bio' in the Netherlands. The enquiry showed that the shipment had been made at Fos-sur-Mer, in the southeast of France, and that the cereals were conventional French ones. Similar trafficking may crop up. It underlines the limitations of the certifying bodies, in charge of controlling the fulfilment of biological agriculture requisites through two annual visits without warning.

There were three certifying organisms in France. The most important one, Ecocert, covered 80 per cent of the market; by mid-1999, it was able to detect pesticide traces in animal feed produced by Central Soya, a neighbour and client of Eurograin. Ecocert then alerted the certifying organism of Eurograin, Afaq-Ascert, which had been controlling this company since 1998. Many officials have recommended the stricter enforcement of biological agriculture requisites, considered as too loose. Fraud and the subsequent mistrust could also explain the relative slump in the consumption of 'bio' products. Another explanation of this decrease is the competition among quality labels as well as the trend toward a more environment-friendly agriculture ('rational' agriculture') that may kidnap the image relating to organic farming.

'Rational' agriculture: "Rational" agriculture's goal is to make a compromise between productivity and environment conservation. In France, the terms of reference of this mode of agricultural production contain 98 points relating to the

improvement of agricultural practices (better use of pesticides and recycling of wastes, improvement of sanitation, precise recording of farmer's practices, better prevention of animal diseases) and upkeep of farm economic profitability.

A key issue of conventional agriculture is the better and more effective use of fertilisers by crops. Increasing the absorption and assimilation of macronutrients such as nitrogen fertilisers would therefore contribute to decreasing the percentage of these fertilisers not used by plants and therefore to drastically reduce soil and water pollution by nitrogen compounds. In this regard the work by Shuichi Yanagisawa and colleagues of the Universities of Okayama and Tokyo, published in the Proceedings of the National Academy of Sciences (PNAS) on 18 May 2004 could be very promising.

The Japanese researchers were able to incorporate into the genome of Arabidopsis thaliana a gene from maize that improves nitrogen assimilation in this crop species. The gene transferred, Dof1, does not only control the synthesis of a protein, but also a series of genetic regulations involved in the constitution of the plant's 'skeleton'. In Arabidopsis thaliana, indeed, Dof1 has modified the expression of several genes: amino-acid concentration has increased, that of glucose was lowered compared with control plants, but not that of sucrose. But the most striking modification concerned the growth of the transgenic plants in an environment with limited amounts of nitrogen: control plants showed symptoms of deficiency such as blemished leaves, while transgenic plants looked normal.

As a follow-up to the experiment on Arabidopsis thaliana, Dof1 has been transferred to potato plants by the Japanese workers; as a result, the amount of amino acids in the genetically modified plants increased. This result is promising, because potatoes are not very efficient in absorbing and assimilating nitrogen compounds; henceforth the need to add such fertilisers to this crop.

Increasing biological nitrogen fixation (instead of nitrogen fertiliser inputs) is also an objective of 'rational' agriculture.

Legumes are able to .x atmospheric nitrogen thanks to symbiotic bacteria living in their root nodules (Rhizobium) and they need much less nitrogen fertilisers than other crop species. An international consortium is carrying out the sequencing of the genome of the annual alfalfa species, Medicago truncatula, considered as a model legume. This genomics work is to be finalised in 2007, but preliminary results already showed that the genomes of legumes were remarkably conserved, i.e. the gene sequence is quite similar among the different species. About 500 genes have been identified as related to the symbiotic relationship between the legume and its rhizobia.

Legume geneticists also hope to unravel the role of the Nod factor, a molecule identified by the researchers of the French National Agricultural Research Institute (INRA) in Toulouse; it has been shown that one-tenth of mg of this substance per hectare was sufficient to raise soybean yields by 11 per cent. Although the transfer of biological nitrogen fixation to cereals is considered a remote possibility because it would involve the transfer of a few hundred genes, the immediate priority is to select the most effective legume species and varieties using genetic methods, so as to raise the percentage of peas, lupins and horsebeans grown in Europe as feed and consequently decrease the dependence of European countries on the imports of soybeans for animal feed (70 per cent of animal feed proteins were imported in 2004) from the USA, Brazil and Argentina. Another objective of the selection of more effective nitrogen-fixing legume species and varieties is to reduce water and soil pollution caused by nitrogen fertilisers (nitrates) , as well as the fossil energy needed to produce these fertilisers: to produce and spray one ton of fertilisers, two tons of oil are needed.

Case of Slow Food: The problems of organic farming may nevertheless be discounted by consumers from Europe or other reach countries on the basis of arguments relating to eating habits, taste and hedonism. Thus, Carlo Petrini, president of the association Slow Food, is setting up in Italy in the heart of a

300-hectare domain (where environment-friendly agriculture methods will be practised) the first world's 'university of taste'. By the end of 2003, some 400 students from the five continents were studying all the aspects of food culture in Piemonte. The ultimate goal is to train specialists that throughout the world will preach the art of good food. Supported by the revenues from a hotel and a wine bank, the university will be completely independent from the food industry. The •16 million needed for the project have been found, especially from private savings.

C. Petrini has built a real counter power over 14 years, starting from a village in Piemonte, but now spreading outside the Italian borders. Slow Food's symbol is the snail—slow and tasty—and its slogan is: 'eat less and eat better'; it has also a web site, publishes a review in four languages and owns an editorial house. About 75,000 persons from Europe and North America adhere to the association and all defend the gastronomic heritage which is, according to them, threatened by the homogenisation of tastes, multinationals and hypermarkets.

In October 2002, 140,000 persons attended the third Congress of Taste, organised by Slow Food in Torino. The Slow Food's prize for the defence of biodiversity was awarded to 13 farmers from Japan, Greece, Guatemala and Guinea. They all had the merit to safeguard a product, such as an old rice variety, Andean vegetables or black piglets. Slow Food is difficult to define: it is a non-governmental organisation, a consumers' association and a gastronomic club, and the whole managed as an advanced enterprise.

Slow Food can claim that it has been successful in achieving its goal of advocating the importance of good food and the emphasis on maintaining and even widening the diversity of food culture, which entails an environment-friendly agriculture that includes organic farming. With 40,000 members in Italy, 9,000 in the USA, 6,000 in Germany and 3,000 in Switzerland, Slow Food is taken seriously by lobbies and big food and beverage companies.

SAFETY OF BIOTECH CROPS

Most crops developed through biotechnology that are on the market today provide farmers with increased convenience and product quality while requiring fewer chemical inputs. According to the USDA Economic Research Service, herbicide- and insect-resistant biotech varieties accounted for about 85 percent of U.S. soybean acreage and 45 percent of corn acreage in 2003. Livestock eat the meal from approximately 70 percent of the soybeans and consume 80 percent of the corn grain and silage grown in the United States, making the livestock industry a major user of biotech crops. Plant breeders are concentrating on enhancing grains or protein sources to produce feedstuffs that will improve feed utilisation, performance, product quality, and health of livestock while reducing production costs and environmental impacts.

A biotech crop is a crop plant that has been genetically engineered using recombinant DNA technology either to promote or to prevent the production of a particular protein, with the objective of introducing or enhancing a desirable characteristic in the plant or seed. Recombinant DNA is DNA formed external to a living cell by joining DNA from two or more different sources. Current biotech crops provide resistance to herbicides, diseases, and insect pests. Herbicide-tolerant crops simplify weed control by allowing growers to apply broad-spectrum herbicides without harming the crop. This permits growers to use fewer herbicides and to shift to herbicides that may be more favourable to the environment. Many growers report that herbicide-tolerant crops make it more convenient to adopt minimum tillage or notill practices that reduce soil erosion. Insect-resistant crops reduce pesticide use and therefore are safer for both workers and the environment.

Crops that are resistant to viral diseases have also been developed using biotechnology, and crops with modified composition or nutritional properties are currently under development. In 2003, 167 million acres of biotech corn, soybeans, cotton, and canola were grown worldwide, and more than 50 biotech varieties have now been approved for use in various countries.

To be grown commercially, biotech crops must first undergo a formal safety assessment by government authorities in the country in which they are grown. In the United States, the Food and Drug Administration (FDA) is responsible for evaluating whether biotech crops are safe for human and animal consumption. The FDA uses the principle of substantial equivalence to assess the safety of biotech-derived crops in food and feed. The biotech crop is compared to its nearest non-biotech counterpart that has a long history of safe consumption. The only significant difference in composition between the two varieties should be the intended presence or absence of the targeted protein that results in the desired characteristic in the biotech plant or seed.

A safety evaluation therefore compares the composition of the biotech crop to that of its conventional counterpart in order to establish substantial equivalence. The FDA also examines the properties of the protein made by the introduced gene in the biotech crop for possible toxicity or allergenicity. The evaluation may include assessments on the performance and health of animals fed with the biotech crop. Further safety tests in addition to compositional analyses and animal feeding studies may be recommended, depending on the characteristics of the introduced trait.

All currently approved biotech crops on the market have been registered for both human food and animal feed use. Prior to approval, feed derived from a biotech crop is analysed to ensure that it has the same digestibility and nutritional composition as feed from conventional varieties of the same crop. Numerous scientific studies evaluating animal performance on biotech feed have been performed on beef cattle, swine, sheep, fish, lactating dairy cows, and chickens. Several reviews summarising the results of studies in many species have concluded that animals fed biotech crops show no differences in performance as compared to animals consuming non-biotech varieties of the same crops. No studies have shown deleterious effects on livestock performance resulting from the consumption of commercially grown, genetically engineered crops.

An unforeseen benefit resulting from feeding of biotech crops has been noted in the case of insect-resistant corn. Molds that often grow at the site of insect damage can produce several fungal toxins, including the deadly fumonisin. Because insectresistant corn sustains less insect damage, it is less susceptible to contamination by these toxins. The resulting corn is safer for both livestock and human consumption.

Animals digest proteins from biotech crops in the same way as they digest proteins from non-biotech crops. Dietary DNA and proteins are degraded during the digestive process. A number of scientific studies indicate that introduced DNA or proteins from biotech crops are not detected in milk, meat, or eggs from animals that consume feed components derived from these crops.

Future Biotech Crops

A new generation of biotech crops designed specifically to benefit the livestock producer is under development. These crops are being engineered with substantial changes in their content of major components (e.g., proteins, amino acids, oils, fatty acids, starches, sugars, fiber) or minor components (e.g., vitamins, minerals, enzymes). As these improved feed crops are designed and intended to be different from non-biotech varieties, they are not expected to be substantially equivalent.

Consequently, their approval process will most likely involve additional assessments of their safety for human and animal consumption, as well as extensive documentation of the performance and health of livestock fed with them. The biotech corn and soybean varieties with increased oil content that are currently under development will provide greater energy intake for beef, swine, and poultry. High-oil feeds based upon natural genetic variation are already available for a limited number of crops, including corn. High-oil corn reduces the amount of feed required for a livestock diet, and this in turn reduces the volume of manure.

Using biotechnology, scientists may be able to enhance the oil content of crops where there is no natural variation for this trait. Furthermore, conventional high-oil crops often have lower yield or protein content than their loweroil counterparts, whereas traits introduced via biotechnology can modify oil accumulation only at specific growth stages and in targeted tissues to minimise such deleterious effects.

Biotech modification of the oil composition of feeds, such as raising the level of oleic acid, may also improve the quality of the resulting animal products for processing and human nutrition. Protein content and feeding quality are also being targeted for improvement in biotech crops. Scientists can modify the protein content of crops either indirectly (by improving nitrogen assimilation) or directly (by modifying key biochemical pathways or introducing proteins with a different amino acid composition).

Researchers have produced soybeans and corn with elevated levels of the essential amino acids lysine, methionine, and threonine, and preliminary reports on livestock performance found that consumption of these crops reduced the need for dietary supplementation with protein and amino acids. Similarly, by elevating the levels of sulfur-containing amino acids in lupins, researchers improved the performance of broiler chickens as well as wool growth and weight gain in sheep. The utilisation and nutritional value of feed crops could be enhanced by increasing their digestibility or decreasing undesirable constituents in feed, which may in turn help to reduce the impact of intensive livestock production on the environment.

Lignin, which is present in the fibrous portion of plant stems, is often indigestible in non-ruminant animals. High lignin content reduces the efficiency of feed utilisation and thereby reduces animal growth. Conventionally bred forage varieties with reduced lignin are available, but they tend to have weaker stems and poor standability in the field. Researchers have developed biotech alfalfa with 20 percent less lignin and 10 percent more cellulose, a combination that makes it more digestible As with any new variety, it will need to be evaluated in field trials to determine whether it has any undesirable characteristics. The ability to modify specific components of fiber biosynthesis may allow scientists to develop reduced-lignin forage that is more digestible and still has the stem strength needed for good field performance.

Phosphorus is an essential nutrient that is critical for animal growth and reproduction and in the formation of the animal skeletal system. Much of the phosphorus in corn and other feedstuffs is unavailable to monogastric (single-stomach) animals such as hogs and poultry because it is bound in a complex form called phytic acid (or phytate). As a result, feeds are often supplemented with phosphorus while the phytate is excreted in the manure. The phytate is eventually broken down by soil microorganisms, and this can result in elevated phosphorus levels in lakes and waterways due to run-off from farms.

Two ways to decrease phytate excretion are to decrease the phytate content of the feed and to add the enzyme phytase to the feed in order to break down the phytate complex. Public and private researchers are working on methods for using biotechnology to increase the availability of phosphorus in animal feeds by enabling crops to express fungal or bacterial phytases that can then be blended into feed rations. If successful, these feeds would help the environment by decreasing the phosphorus content of animal waste.

9

Crop Biotechnology and Food Safety

For many years the practical difficulties of obtaining meaningful information from conventional toxicology studies on the safety of whole foods have been well recognised. The limitations of conventional toxicological studies became particularly apparent when animal feeding studies were used to assess the safety of irradiated foods. Animal studies are a major element in the safety assessment of many compounds such as pesticides, pharmaceuticals, industrial chemicals and food additives. In most cases however, the test substance is well characterised, of known purity, of no particular nutritional value and human exposure is generally low.

It is therefore relatively straightforward to feed such compounds to animals at a range of doses, some several orders of magnitude greater than the expected human exposure levels, in order to identify any potential adverse health effects of importance to humans. In this way it is possible, in most cases, to determine levels of exposure at which adverse effects are not observed, and so set safe upper limits by the application of appropriate safety factors. By contrast, foods are complex mixtures of compounds characterised by wide variation in composition and nutritional value. Due to their bulk and effect on satiety they can usually only be fed to animals at low multiples of the amounts that might be present in the human diet.

In addition, a key factor to consider in conducting animal studies on foods is the nutritional value and balance of the diets used, to try to avoid the induction of adverse effects which are not related directly to the material itself. Detecting any potential adverse effects and relating these conclusively to an individual characteristic of the food can therefore be extremely difficult. Another consideration in deciding the need for animal studies is whether it is appropriate to subject experimental animals to such a study if it is unlikely to give rise to meaningful information. In practice, very few foods consumed today have been subject to any toxicological studies, yet they are generally accepted as being safe to eat. In developing a methodology for the safety assessment of new foods, it was essential to establish a benchmark definition of safe food. This was taken up by OECD in 1991 who said that food is considered safe if there is reasonable certainty that no harm will result from its consumption under anticipated conditions of use.

The difficulties of applying traditional toxicological testing and risk assessment procedures to whole foods, meant that an alternative approach was required for the safety assessment of genetically modified foods. This led to development of the concept of substantial equivalence. This approach acknowledges that the goal of the assessment is not establishing absolute safety but to consider whether the genetically modified food is as safe as its traditional counterpart, where such a counterpart exists.

The Consultation agreed that the practical difficulties already identified in relation to the application of conventional toxicology studies to whole food preclude their use as a routine safety assessment technique for genetically modified foods. The Consultation also recognised that the use of toxicology studies could not be justified from an animal welfare perspective where it was unlikely to result in meaningful information. In addition, the Consultation noted that the concept of substantial equivalence was a good example of an approach to reduce the use of animals in toxicology studies by refining safety assessment techniques and replacing animal models with alternatives.

The concept of substantial equivalence was developed as a practical approach to the safety assessment of genetically modified foods. The Consultation agreed that substantial equivalence should be seen as a key step in the safety assessment process. The application of the concept is not a safety assessment in itself; it does not characterise the hazard, rather it is used to structure the safety assessment of a genetically modified food relative to its conventional counterpart. As a starting point, the genetically modified organism (plant, micro-organism or animal), and/or foods derived from it, is compared with its closest traditional counterpart in order to identify any intended and unintended differences which then become the focus of the safety assessment.

Data for comparison should be obtained using validated methods and analysed using appropriate statistical techniques. The comparative approach should take into account agronomic, genetic and chemical aspects and only when all of these have been considered can an objective assessment of safety be made. The type and extent of further studies depend on the nature of the differences and whether or not they are well characterised. Studies should be carried out in accordance with good laboratory practise.

Food Safety Assessment

Several international organisations have already addressed the issues associated with the safety assessment of novel foods and, genetically modified plants and micro-organisms (WHO, 1991; OECD, 1993; WHO, 1995; FAO, 1996; ILSI, 1996; Commission of the European Communities, 1997). It is generally agreed that such an assessment requires an integrated and stepwise, case-by-case approach and some authorities have developed decision trees to assist in determining the extent of testing required in specific cases (SCF, 1997; UK ACNFP, 1995). This approach is useful in determining appropriate safety assessment strategies.

The safety assessment of a genetically modified food is directed by the results of a comparison between the genetically modified food and its conventional counterpart. It follows a stepwise process aided by a series of structured questions. Factors taken into account in the safety assessment include:

— identity;

— source;

— composition;

— effects of processing/cooking;

— transformation process;

— the recombinant DNA (e.g. stability of insertion, potential for gene transfer);

— protein expression product of the novel DNA;

— effects on function;

— potential toxicity;

— potential allergenicity;

— possible secondary effects from gene expression or the disruption of the host DNA or metabolic pathways, including composition of critical macro-, micro-nutrients, anti-nutrients, endogenous toxicants, allergens, and physiologically active substances; and,

— potential intake and dietary impact of the introduction of the genetically modified food.

The above factors are particularly pertinent to the assessment of foods derived from genetically modified plants. When assessing the safety of foods derived from genetically modified animals and micro-organisms, other factors may need to be taken into account on a case-by-case basis.

If the characterisation of the food indicates that the available data are insufficient for a thorough safety assessment, animal testing may be deemed necessary. This would particularly be the case if the food were expected to make a significant dietary contribution, if there is no history of consumption of the novel gene product or if the modification affects several metabolic pathways. In the situation where the genetically modified food differs from the traditional counterpart by the presence of one or a few new genes and their products, it may be possible to isolate and study these in a manner analogous to conventional toxicity testing of food additives.

However it is essential to ensure that the material tested is biochemically and functionally equivalent to that produced in the genetically modified food. This provides the possibility of increasing the sensitivity of toxicity tests compared with that possible if the products of the genetically modified plants had been fed directly and avoids some of the artefacts that can occur in toxicity tests on whole foods. However, this strategy is only applicable if the preceding detailed analysis does not reveal significant changes other than those expected. Otherwise testing of the whole food may be required. When animal testing is conducted on the whole food, it should generally be on the food as consumed by humans.

The type of animal study would need to be considered on a case by case basis. In addition to investigating potential toxicological effects, animal studies may also be required if the genetic modification directly or indirectly affects the content or bioavailability of nutrients. Where toxicological studies are considered necessary to assess the safety of long term consumption of a food in the diet, it is generally considered that a sub-chronic study of 90-days duration is the minimum requirement to demonstrate the safety of repeated consumption of a food in the diet.

This may need to be preceded by a pilot study of short duration to ensure that the diet is palatable to the test species and that the levels of incorporation of the test article are appropriate, e.g. the control diet containing the equivalent level of the comparator does not produce effects, as a result of normal levels of natural toxicants present in traditional foods accepted as safe. The highest dose level used in any animal study should be the maximum achievable without causing nutritional imbalance while the lowest level used should be comparable to the anticipated human intake.

The need for additional toxicological tests should be considered on a case-by-case basis taking into account the results of the 90-day study and other studies. For example, proliferative changes in tissues during the 90-day study may indicate the need for a longer-term toxicity study. Conventional toxicological tests are of limited value in assessing whole foods, including genetically modified foods. Based on the maximum levels of the whole food that can be incorporated into experimental diets as indicated previously, a margin of safety may be estimated based on the absence or nature of adverse effects and likely human exposure.

Improved experimental designs should take into account the need for nutritionally adequate animal diets, avoiding some of the inappropriate testing of foods or products. It has been suggested that the use of biomarkers of early effects might increase diagnostic value and sensitivity of toxicity tests on foods. However, it will be necessary not to confuse adaptive and toxic effects in applying this approach.

In achieving the objective of conferring a specific target trait (intended effect) to the host organism by the insertion of defined DNA sequences, additional traits could, theoretically, be acquired or existing traits lost (unintended effects). The assessment of genetically modified foods involves methods to detect such unintended effects and procedures to evaluate their biological relevance and their impact on food safety. Unintended effects may be due to factors such as random insertion events which might result in disruption of existing genes, modifications of protein expression or formation of new metabolites. The expression of enzymes at high levels may give rise to secondary biochemical effects, e.g. an altered metabolic flux resulting in changed metabolite patterns.

The potential occurrence of unintended effects is not specific to the use of recombinant DNA techniques. Rather, it is an inherent and general phenomenon that can occur in conventional breeding. One of the approaches adopted to cope with this problem is to select/ discard plants with unusual and undesired phenotypic and agronomic parameters at an early stage of the plant variety development. The practice of consecutive back-crossing is also a common procedure used to eliminate unintended effects.

Unintended effects due to genetic modification may be subdivided into two groups: those which are "predictable" based on metabolic connections to the intended effect or knowledge of the site of insertion and those which are "unexpected". Due to the increased precision of genetic modification compared to conventional breeding, it may become easier to predict pathways likely to be influenced by unintended effects. The comparator used to detect unintended effects should ideally be the near isogenic parental line grown under identical conditions. In practice, this may not be feasible at all times, in which case a line as close as possible should be chosen.

The resulting natural variation should be taken into account in assessing the statistical significance of the unintended effect. Where statistically significant unintended differences are observed, their biological significance should be assessed. This may be assisted by knowledge of the mechanisms leading to the changes. In order to assess the biological and safety relevance of an unintended effect, data on the genetically modified plant should be compared to data on other conventional varieties and literature data. If the differences exceed natural variations in traditional food crops, further assessment is required. Present approaches to assess possible unintended effects are based, in part, on the analysis of specific components (targeted approach).

In order to increase the probability of detecting unintended effects, profiling techniques are considered as useful alternatives (non-targeted approach). Profiling techniques are used at different level e.g. genomics, proteomics and metabolomics, and may contribute to the detection of differences in a more extensive way than targeted chemical analysis. However, they are not yet fully developed and validated and have certain limitations. In the future, genetic modifications of plants are likely to be more complex perhaps involving multiple between-species transfers and this may lead to an increased chance of unintended effects. Where differences are observed using profiling techniques, the possible implications of the differences with respect to health need to be considered.

Substantial Equivalence

The Consultation acknowledged that the concept of substantial equivalence had attracted criticism. This criticism relates, in part, to the mistaken perception that the determination of substantial equivalence was the end point of a safety assessment rather than the starting point. Having considered the way in which the concept of substantial equivalence is currently used, and the possible use of alternative strategies, the Consultation concluded that application of the substantial equivalence concept contributes to a robust safety assessment framework.

The Consultation was satisfied with the approach used to assess the safety of the genetically modified foods that have been approved for commercial use. It was agreed that communication of the principles involved in safety assessment could be improved. The Consultation concluded that the key message to be conveyed is that substantial equivalence is a concept used to identify similarities and differences between the genetically modified food and a comparator with a history of safe food use which subsequently guides the safety assessment process.

The Consultation reiterated that a consideration of compositional changes was not the sole basis for determining safety. Safety can only be determined when the results of all aspects under comparison are integrated. It was recognised that whole foods do not lend themselves to the standard safety evaluation principles used for food additives and other chemicals and that a quantitative assessment of risk of individual whole foods from whatever source cannot be achieved. The Consultation agreed that assessing safety relative to existing foods offered the best means of assessing the safety of genetically modified foods.

The Consultation considered the issue of long term effects from the consumption of genetically modified foods and noted that very little is known about the potential long term effects of any foods. In many cases, this is further confounded by wide genetic variability in the population, such that some individuals may have a greater predisposition to food-related effects. Experimental studies, such as randomised controlled trials (RCTs), if properly designed and conducted, could be used to investigate the medium/long term effects of any foods, including genetically modified foods. Such studies could provide additional evidence for human safety, but would be difficult to conduct. In this respect, it is also important to recognise the wide variation in diets and dietary components from day to day and year to year.

The Consultation was of the view that there were presently no alternative strategies that would provide a better assurance of safety for genetically modified foods than the appropriate use of the concept of substantial equivalence. Nevertheless, it was agreed that some aspects of the steps in the safety assessment process could be refined to keep abreast of developments in genetic modification technology. New methodologies, such as profiling techniques, offer the means of providing a more detailed analytical comparison. However, it was recognised that much more developmental work was necessary before such methods could be validated.

Nutritional Issues

The variety of foods consumed by humans has changed greatly

over the centuries, altering the balance of nutrients in the diet. Plant breeding by traditional methods, mutations and recombinant DNA technique can be used to alter nutritional quality and functional traits. Thirty years ago, traditional breeding was utilised to identify and select rapeseed plants free of the nutritionally undesirable fatty acid, erucic acid. The resulting plant, canola, produced an oil that has a more desirable fatty acid profile.

Canola oil now makes up a significant proportion of the daily lipid intake of the consumers in most of the developed nations of the world. More recently, mutational breeding has been used to genetically modify flax plants to produce edible oils instead of the traditional high linolenic industrial oil. The introduction of new foods and the growing interest in functional foods also have the potential to modify the food supply. The availability of recombinant DNA techniques provides the opportunity to develop foods that help optimise health status. The major difference with recombinant DNA techniques is the ability to introduce different nutrient profiles with greater speed and precision. Genetic modification of plants now under cultivation have been directed towards agronomic enhancement.

Nutritional changes may have a more profound impact on the health of the population. At present, there are no foods derived from genetically modified plants modified to enhance nutrition in the commercial market. However, there are several plants with altered nutrient composition being developed using recombinant DNA technology. These have been designed to modify nutrient composition and levels or change the functionality of a product. An example of the latter is potato tubers containing increased amounts of starch that is distributed more uniformly, resulting in more efficient processing, lower fat absorption and improved texture. It is anticipated that other genetically modified plants will be developed with nutritional characteristics targeting major health problems.

There are a number of examples of genetically modified plants from which foods are produced with the objective of improving health and enhancing food functionality. These include two examples of genetically modified rice varieties, one in which beta-carotene was produced and another in which an undesirable component for sake brewing (glutelin) was decreased. Other examples were oil seeds in which the fatty acid profile was changed either by traditional mutational techniques or recombinant DNA technique.

Recently, canola and soy oils with combined low levels of saturated fatty acids and increased oleic acid, have been produced with the objective of lowering total and low density lipoprotein (LDL) cholesterol levels, one of the risk factors for cardiovascular disease, while at the same time enhancing the functionality of the oil. The Consultation heard details of the recently reported "golden rice" which was specifically designed to target Vitamin A deficiency, a cause of blindness among people living in developing countries. These examples highlight the potential of foods with modified nutritional profiles to reduce the incidence of nutrition-related conditions or diseases.

Traditional plant breeding techniques of intra and inter-species crossing and mutation are designed to create genetic variation upon which selection of the most desired genotype is the expected outcome. All plant breeding procedures can produce unexpected effects. Low glutelin genetically modified rice, created using an anti-sense technique, signifies improvements in rice storage proteins for commercial sake brewing. The decrease in glutelin levels was however associated with an unintended increase in levels of prolamins. This illustrated that a targeted change in the level of a specific protein can result in other proteins being affected.

The change in prolamin levels did not affect the industrial application but could affect nutritional quality and allergenic potential if the rice were used as a food. Other examples were given regarding soybean and rice showing that genetically modified soy with increased lysine showed an unexpected decrease in oil content, and the genetically modified "golden rice" designed to express betacarotene unexpectedly accumulated xanthophylls. The use of tissue-specific promoters was suggested as a means to limit the number and extent of unexpected effects. In the case of the low glutelin rice, the change in prolamins would not be detected by standard nutritional analyses such as total protein and amino acid profiles. This was only observed following sodium dodecyl sulphate (SDS) gel electrophoresis.

The unexpected finding of xanthophylls in the beta-carotene-enhanced genetically modified rice would not have been apparent from standard nutritional analyses. This difference was observed following high pressure liquid chromatography (HPLC) analyses for carotenoids. Thus, it is important that appropriate analyses of nutrients should be carried out to identify unexpected changes. The ability to change nutrient levels in crop plants through plant breeding, including the use of recombinant DNA techniques, has the potential to result in broad changes in at least two ways: (1) the intended modification in plant constituents could change the overall nutrient profile of the plant product and this change could affect nutritional status of the individual, (2) in addition, unexpected alterations in nutrients could also affect nutrient profiles of the product and nutritional status of people.

Although the genetically modified plant components may be assessed as safe individually, the impact of the change on the overall nutrient profile must be determined. Because changes in individual nutrients could affect a number of plant processes and nutritional outcomes, it is recommended that integration of nutritional and toxicological expertise needed for the evaluation of genetically modified foods be encouraged and facilitated. Consideration should be given to assessing the potential health impact resulting from changes in nutrient profile arising from all types of plant breeding.

Examples of the use of mutational breeding to alter the nutritional characteristics of plants include the modification of flaxseed oil from a high linolenic industrial oil to a high linolenic oil similar to corn oil in its fatty acid composition, and genetically modified soy and canola plants being developed that produce a high oleic acid (80-90%) oil that also displays very low levels of saturated and polyunsaturated fatty acids. It will be important to determine if the overall nutrient profile of a genetically modified food has been changed and if dietary intake patterns are altered by the introduction of foods from genetically modified plants.

The introduction of a significant nutritional change in a food may require post-market assessment to determine whether the overall diet has been altered and to what degree, before an assessment of the impact on nutritional status can be made. It is important to ascertain to what extent the modified nutrient is bioavailable and remains stable with time, processing, and storage. For example, the question was raised as to what extent carotenoids in the genetically modified rice remained stable under storage conditions encountered in the developing countries.

Where additional assurance of safety is sought, analytical methods traditionally applied in the evaluation of food constituents such as total protein, fat, ash, fibre and micronutrients may need to be augmented with additional analyses tc identify unexpected effects and altered nutrient profiles and bioavailability which may impact on dietary intake and health. Because of the potential for broad changes in nutrient levels and interactions with other nutrients and unexpected effects, it may be necessary in certain instances to undertake feeding studies in animals to determine outcomes that result from changes in nutrient profiles and nutrient bioavailability.

Nutritional modifications which are within the normal range of nutrient variation might require a less extensive evaluation than those outside normal ranges. Genetically modified foods have the potential to improve the nutritional status of individuals and provide products with enhanced functionality for populations in developed and developing countries. The major issues relate to possible nutritional imbalances and the introduction of unexpected alterations in nutrients and other compounds.

The change in nutrient levels in a particular crop plant may impact overall dietary intake. In such cases, it would be important to monitor changes in nutrient levels and bioavailability in such foods and evaluate their potential effect on nutritional and health status of consumers. However, an assessment of the impact on nutritional status of consumers is important for all significant dietary changes and not specific to the introduction of genetically modified foods.

Food Safety Issues

There are numerous issues, that frequently raised with regard to the safety of genetically modified foods. These issues include the potential for gene transfer from genetically modified plants to gut microflora and mammalian cells, the safety of antibiotic resistance genes as markers for the selection of genetically modified plants, and the assessment of the potential allergenicity of genetically modified foods that may be caused by the presence in these foods of novel gene products.

Gene Transfer

In humans, dietary intakes of RNA and DNA vary widely but are typically in the range from 0.1 to 1.0 g per day. Any concerns over the presence of novel DNA in a genetically modified food consumed in the human diet must take into consideration that this DNA would represent less than 1/250,000 of the total amount of DNA consumed. In view of this and the digestibility of dietary DNA, the probability of transfer of genes from genetically modified plants to mammalian cells is extremely low. It is nevertheless necessary to examine this possibility and the consequences of such transfer if it were to occur. The transfer of plant DNA into microbial or mammalian cells under normal circumstances of dietary exposure would require all of the following events to occur:

— the relevant gene(s) in the plant DNA would have to be released, probably as linear fragments;

— the gene(s) would have to survive nucleases in the plant and in the gastrointestinal tract;

— the gene(s) would have to compete for uptake with dietary DNA;

— the recipient bacteria or mammalian cells would have to be competent for transformation and the gene(s) would have to survive their restriction enzymes; and

— the gene(s) would have to be inserted into the host DNA by rare repair or recombination events.

There have been numerous experiments aimed at evaluating the possibility of transfer of plant DNA to microbes and mammalian cells. To date, there are no reports that marker genes in plant DNA transfer to these cells. Results of model experiments in which mice were orally administered high doses of bacterially derived DNA indicated apparent incorporation of the test DNA fragments into bacterial and mouse cells. The report contrasts with other reports where no transfer or only a low frequency of transfer was observed. The significance of the observations of Schubbert et al. have been seriously questioned. It was concluded that the data do not demonstrate that plant DNA can be transferred to and stably maintained in mammalian cells.

There is additionally no evidence that intact genes from plants can be transferred to and be expressed in mammalian cells. It should be noted that the vast majority of known bacteria are not naturally transformable and there is as yet no evidence of transfer to and expression of plant genes in bacteria under natural conditions. Transfer has been observed under laboratory conditions, but only if homologous recombination is possible. It should be noted that inserted gene sequences in genetically modified plants show in many cases homology with prokaryotic genes.

The Consultation is aware of the study being undertaken whereby chickens and sheep are being fed genetically modified maize, and bacteria in the normal flora of the gastrointestinal tract are being tested for DNA uptake. Should horizontal gene transfer from a genetically modified plant to bacteria occur, the gene (e.g. an antibiotic resistance gene) may alter the fitness of the recipient cell. A reduction in fitness may not provide sufficient selective pressure to eliminate the gene or gene fragment from the gene pool.

The presence of this DNA in the cell population could then serve as a genetic reserve for the evolution of the recipient species. The available knowledge on bacteria is derived from bacteria that can be cultured and readily analysed. Bacteria that are susceptible neither to culture nor identification represent a significant proportion of existing microflora. Therefore, without available knowledge of these bacteria, it is not possible to assess the possibility, probability or consequences of their acquisition of genes or gene fragments. The consequences of uptake of plant DNA by mammalian cells are different from those of uptake by bacteria because existing data indicate that such DNA is not transmitted via the germline.

The extent to which cells containing exogenous DNA are phagocytosed is not yet clear. Neither is it clear that the incorporated DNA is stably maintained and replicated in somatic cells. Mammalian cells would be similarly affected by uptake of exogenously derived DNA regardless of its source. The most important consideration with respect to horizontal gene transfer is the consequence of a gene being transferred from a genetically modified plant and expressed in recipient cells. Therefore data on the possible extent of such transfer will be needed as part of the safety assessment when the nature of the transferred gene(s) is such that, if transfer were to occur, it would give rise to health concerns.

The Consultation noted that the antibiotic resistance markers currently used in genetically modified plants have been previously reviewed for safety. There is no evidence that the markers currently in use pose a health risk to humans or domestic animals. Nevertheless, with the variable gene transfer frequencies noted in current literature, the transfer and expression of a functional antibiotic resistance gene to recipient cells, while remote, cannot be ignored. If the recipient cell is subjected to selection from therapeutic use of the antibiotic, proliferation of a drug resistant cell population could compromise the effectiveness of the drug. This directs attention to the more important considerations: whether there are already prevailing high levels of culturable bacteria resistant to that antibiotic, whether that antibiotic is, or could be, clinically important, and whether there are alternative effective therapies.

For certain antibiotic resistance genes currently in use in genetically modified plants, the available data suggest that the consequences of horizontal gene transfer will be unlikely to pose a significant threat to the current therapeutic use of the respective drugs. With other genes that confer resistance to drugs that are important in specific medical use, or to drugs that have limited alternative therapies, the possibility of transfer and expression of these genes is a risk that warrants their avoidance in the genomes of widely disseminated genetically modified organisms and foods and food ingredients.

A number of methods are available to genetically modify plants without incorporation of an antibiotic resistance gene in the commercial product. These methods include removing the gene after successful gene transfer, or using alternative marker genes for genetic transformation. If alternative marker genes are used, they also would need to be evaluated for safety.

In addition, it is recognised that further technical development of these or additional methods may be necessary for practical transformation of certain plant species. In future developments, the Consultation encourages the use of alternative transformation technologies, if available and demonstrated to be safe, that do not result in antibiotic resistance genes in genetically modified foods. If further development of alternative technologies is required, additional research should be strongly encouraged.

ALLERGENICITY

Food allergies are adverse reactions to an otherwise harmless food or food component that involves an abnormal response of the body's immune system to specific protein(s) in foods. True food allergies may involve several types of immunological responses. The most common type of food allergies are mediated by allergen-specific immunoglobulin E (IgE) antibodies. IgE-mediated reactions are known as immediate hypersensitivity reactions because symptoms occur within minutes to a few hours after ingestion of the offending food.

IgE-mediated reactions can occur to pollens, mould spores, animal danders, insect venoms and other environmental stimuli as well as foods. IgE-mediated reactions affect perhaps 10-25% of the population in developed countries, although food allergies represent a small fraction of all allergic diseases. IgE-mediated food allergies affect less than 2.5% of the population in developed countries. Infants and young children are more commonly affected by IgE-mediated food allergies than adults; the prevalence among infants under the age of 3 may be as high as 5-8%. True food allergies also include cell-mediated reactions which involve sensitised tissue-bound lymphocytes rather than antibodies.

In cell-mediated reactions, the onset of symptoms occurs more than 8 hours after ingestion of the offending food. The role of foods in cell-mediated reactions remains uncertain but, celiac disease, also known as glutensensitive enteropathy, affects 1 in every 300 to 3000 individuals in the population depending upon the specific geographic region. The Codex Alimentarius Commission has adopted a list of the most common allergenic foods associated with IgE-mediated reactions on a world-wide basis that includes peanuts, soybeans, milk, eggs, fish, crustacea, wheat, and tree nuts. These commonly allergenic foods account for over 90% of all moderate to severe allergic reactions to foods, although an extensive literature search has revealed more that 160 foods associated with sporadic allergic reactions.

Allergic reactions to fresh fruits and vegetables, the so-called oral allergy syndrome, are also rather common, but these foods are not included on the Codex Alimentarius Commission list because the symptoms are typically mild and confined to the oropharyngeal region and the allergens are unstable to heating and digestion. The list established by the Codex Alimentarius Commission also includes gluten-containing cereals (wheat, rye, barley, oats and spelt) that are implicated in the etiology of gluten-sensitive enteropathy. The symptoms of IgE-mediated food allergies can range from mild to severe and lifethreatening.

Individuals display different thresholds for the offending food, but the most sensitive food-allergic individuals will experience reactions from exposure to trace quantities of the offending food. Life-threatening reactions usually involve exposure to larger doses of the offending food. Gluten-sensitive enteropathy is a mal-absorbtion syndrome characterised by body wasting, anaemia, diarrhoea, and bone pain along with other symptoms. The threshold dose needed to provoke the symptoms of gluten-sensitive enteropathy are unknown, but also thought to be quite low.

Both IgE-mediated food allergies and gluten-sensitive enteropathy are treated with specific avoidance diets. Since in both cases, the threshold dose is quite low, great care must be taken in the construction of safe and effective avoidance diets. Almost all food allergens are proteins, although the possibility exists that other food components may act as haptens. Similarly, prolamin proteins from wheat, rye, barley, etc. are involved in the elicitation of gluten-sensitive enteropathy. While the crops from which staple foods are derived contain ten of thousands of different proteins, relatively few are allergenic.

The distribution of these proteins varies in different parts of the plant and can be influenced by environmental factors such as climate and disease stress. Conventional breeding introduces additional protein diversity into the food supply. However, variations in the protein composition of our diets brought about through conventional crop improvement practices have had little, if any, effect on the allergenic potential of our major foods. In contrast, altered dietary preferences can have significant implications for the development of food allergies.

For example, allergy to peanut (groundnut) occurs at a significant frequency in North America and Western Europe but not in other countries where peanuts are less commonly eaten. Also, recent food introductions such as kiwi fruit have proven to be additional sources of food allergens. These observations provide confidence that there are not a large number of potential allergens in the food supply, but show that new allergenic foods are sometimes introduced into the marketplace. Because of the above, a clear need exists to pay particular attention to allergenicity when assessing the safety of foods produced through genetic modification.

Novel Proteins in Genetically Modified Foods

In 1996, the International Food Biotechnology Council and the Allergy and Immunology Institute of the International Life Sciences Institute developed a decision-tree approach. This allergy assessment strategy has been widely adopted by the agricultural biotechnology industry. This strategy focuses on the source of the gene, the sequence homology of the newly introduced protein to known allergens, the immunochemical binding of the newly introduced protein with IgE from the blood serum of individuals with known allergies to the source of the transferred genetic material, and the physicochemical properties of the newly introduced protein.

Since genetically modified foods usually contain novel proteins, their safety should include an assessment of the allergenicity of such novel proteins. The current decision-tree approach requires the examination of a number of parameters which are common to many food allergens. These characteristics facilitate the identification of potentially allergenic gene products, although no single criterion is sufficient to confirm allergenicity or the lack thereof. The relevant criteria used in the current decision tree include:

— *Source of the transferred genetic material*: Particular caution must be exercised if the source of this material contains known allergens.

— *Sequence homology*: The amino acid sequence of many allergens is readily available.

— *Immunoreactivity of the newly introduced protein*: If the novel protein is derived from a known allergenic source or if it has sequence homology with a known allergen, then the reactivity of this novel protein with IgE from the blood serum of appropriate allergic individuals is determined.

— *Effect of pH and/or digestion*: Most allergens are resistant to gastric acidity and to digestive proteases.

— *Heat or processing stability*: Labile allergens in foods that are eaten cooked or undergo other processing before consumption are of less concern.

When the genetically modified food contains genes selected from sources with known allergenic effects, then it must be assumed that the novel gene product is allergenic unless proven otherwise. The current decision-tree approach which advocates the assessment of the binding of the novel protein with IgE from the blood serum of individuals who are allergic to the source of the donor genetic material followed, if necessary by skin testing and blinded oral food challenges, was considered adequate and essential.

The assessment of any unintended effects on the allergenicity of the host material after a genetic modification with genes from other sources, whether allergenic or not, was not considered necessary except in circumstances where the genetic modification could be predicted to alter the protein content of the host product significantly. When the genetically modified food contain genes from sources with no history of allergenicity, the current decision-tree approach relies primarily upon sequence homology comparisons to known allergens and the stability of the novel protein to digestion and processing.

It is widely recognised that these two criteria alone may not be sufficient to assess the potential allergenicity of genetically modified foods containing genes from sources with no history of allergenicity. The current criteria used to determine significant sequence similarity, a match of at least eight contiguous, identical amino acids has been criticised. Suggestions have been made that sequence similarity should instead require a match of a smaller number of contiguous, identical amino acids, perhaps as few as four amino acids.

The use of a match of eight contiguous, identical amino acids appears to have some relevance based upon the minimum peptide length for a T cell-binding epitope. Also, it is recognised that the criterion cannot identify discontinuous or conformational epitopes that depend upon the tertiary structure of the protein. However, the stability of food allergens to heat processing argues for greater significance of linear, continuous epitopes for these particular allergens. International scientific consensus should be sought on the use of sequence homology in the assessment of the allergenicity of genetically modified foods.

The use of digestive stability appears to be a rather useful criterion in the assessment of the allergenicity of genetically modified foods. Simulated gastric and intestinal digestive models of mammalian digestion have been used to assess the digestive stability of known food allergens and proteins introduced into foods through genetic modification. While the usefulness of this criterion is apparent, consensus is needed on the ideal protocols for assessment of digestive stability. It is recognised that novel proteins may exist that are stable to digestion but will not become allergens.

Additional testing is needed to assess the allergenic potential of such proteins. The desirability of the development of additional tests to assess the allergenic potential of foods containing genes from sources with no history of allergenicity has been widely expressed. Two additional tests seem to show some promise for addition to the decision-tree approach. The level and site of expression of the novel protein is an important component of the assessment of allergenicity. Novel proteins expressed at comparatively low amounts in the food would have limited potential for allergic sensitisation.

Major food allergens are usually major proteins in commonly consumed foods. Thus, greater scrutiny should occur with genetically modified foods containing novel proteins at significant levels in the product. New proteins expressed in non-edible portions of plants are not a concern in terms of food allergy. Consideration of the function of the novel protein should also form part of the decisiontree assessment of allergenic potential. Certain classes of proteins are well known allergens. For example, the 2S, high-methionine albumins from Brazil nut, walnut, sunflower seed and mustard are major allergens from those sources. Thus, other 2S, high-methionine albumins should be scrutinised very carefully for allergenic potential.

Many of the pathogenesis-related proteins of plants display allergenic activity, therefore, the entire class of proteins would also merit close examination. International consensus should be sought on a list of functional proteins with allergenic potential. Certainly, other proteins not on such a list must also be evaluated but this aspect could be a useful part of an overall assessment strategy.

Unfortunately, reliable, well validated animal models for the assessment of the allergenicity of genetically modified foods do not presently exist, although further research on the development of animal models is encouraged. The novel proteins present in genetically modified foods should also be evaluated for any possible role in the elicitation of gluten-sensitive enteropathy.

Clearly, if the desired gene is obtained from wheat, rye, barley, oats, or related cereal grains, the possible role of the novel protein in provoking gluten-sensitive enteropathy must be carefully considered. Furthermore, if genetic modifications are conducted on these cereal grains, possible unintended effects on the gluten proteins should be considered. International consensus should be sought on an appropriate decision-tree approach to the assessment of the role of genetically modified foods and their novel proteins in gluten-sensitive enteropathy.

Genetic modification offers the opportunity to decrease or eliminate the protein allergens that occur naturally in specific foods. An example is the development of a genetically modified rice variety developed through anti-sense technology, which dramatically reduced levels of the major rice allergen.

10

ECOLOGICAL IMPACTS OF CROP BIOTECHNOLOGY

Transgenic crops are the main products of agricultural biotechnology. Transnational corporations (TNCs) such as Monsanto, DuPont, and Novartis, the main proponents of biotechnology, argue that carefully planned introduction of these crops should reduce or even eliminate the enormous crop losses due to weeds, insect pests, and pathogens. In fact, they argue that the use of such crops will have added beneficial effects on the environment by significantly reducing the use of agrochemicals.

However, ecological theory predicts that as long as transgenic crops follow closely the pesticide paradigm prevalent in modern agriculture, such biotechnological products will do nothing but reinforce the pesticide treadmill in agroecosystems, thus legitimizing the concerns that many environmentalists and some scientists have expressed regarding the possible environmental risks of genetically engineered organisms. In fact, there are several widely accepted environmental drawbacks associated with the rapid deployment and widespread commercialization of such crops in large monocultures, including:

— the spread of transgenes to related weeds or conspecifics via crop-weed hybridization

— reduction of the fitness of non-target organisms through the acquisition of transgenic traits via hybridization

— the rapid evolution of resistance of insect pests such as Lepidoptera to Bt

— accumulation of the insecticidal Bt toxin, which remains active in the soil after the crop is ploughed under and binds tightly to clays and humic acids;

— disruption of natural control of insect pests through intertrophic-level effects of the Bt toxin on predators

— unanticipated effects on non-target herbivorous insects (i.e., monarch butterflies) through deposition of transgenic pollen on foliage of surrounding wild vegetation14

— vector-mediated horizontal gene transfer and recombination to create new pathogenic organisms

Agrodiversity and Farmers

The spread of transgenic crops threatens crop diversity by promoting monocultures which leads to environmental simplification and genetic erosion. History has repeatedly shown that uniformity characterizing agricultural areas sown to a smaller number of varieties is a source of increased risk for farmers, as the genetically homogeneous fields may be more vulnerable to disease and pest attack. Several people think that HRCs and Bt crops have been a poor choice of traits to feature the technology, given predicted environmental problems and the issue of resistance evolution.

In fact, there is enough evidence to suggest that both these types of crops are not really needed to address the problems they were designed to solve. On the contrary, they tend to reduce the pest management options available to farmers. There are many alternative approaches, (e.g., rotations, polycultures, cover crops, biological control, etc.) that farmers can use to effectively regulate the insect and weed populations that are being targeted by the biotechnology industry. To the extent that transgenic crops further entrench the current monocultural system, they impede farmers from using a plethora of alternative methods.

Insect Resistant Crops

According to the industry, the promise of transgenic crops inserted with Bt genes is the replacement of synthetic insecticides now used to control insect pests. Since most crops have a diversity of insect pests, insecticides will still have to be applied to control pests other than Lepidoptera not susceptible to the endotoxin expressed by the crop. On the other hand, several Lepidoptera species have been reported to develop resistance to Bt toxin in both field and laboratory tests, suggesting that major resistance problems are likely to develop in Bt crops which through the continuous expression of the toxin create a strong selection pressure.

Given that a diversity of different Bt-toxin genes have been isolated, biotechnologists argue that if resistance develops alternative forms of Bt toxin can be used. However, because insects are likely to develop multiple resistance or cross-resistance, such strategy is also doomed to fail. Others, borrowing from past experience with pesticides, have proposed resistance management plans with transgenic crops, such as the use of seed mixtures and refuges. In addition to requiring the difficult goal of regional coordination between farmers, refuges have met with poor success for chemical pesticides, due to the fact that insect populations are not constrained within closed systems, and incoming insects are exposed to lower doses of the toxin as the pesticide degrades.

By keeping pest populations at extremely low levels, Bt crops can starve natural enemies as these beneficial insects need a small amount of prey to survive in the agroecosystem. Parasites would be most affected because they are more dependent on live hosts for development and survival, whereas some predators could theoretically thrive on dead or dying prey.

Natural enemies could also be affected directly through inter-trophic level interactions. Evidence from studies conducted in Scotland suggest that aphids were capable of sequestering the toxin from Bt crops and transferring it to its coccinellid predators, in turn affecting reproduction and longevity of the beneficial beetles. Sequestration of plant allelochemicals by herbivores which then affect parasitoid performance is not uncommon. The potential of Bt toxins moving through food chains poses serious implications for natural biocontrol in agroecosystems.

Bt toxins can be incorporated into the soil through leaf materials, where they may persist for 2-3 months, resisting degradation by binding to soy clay particles while maintaining toxin activity. Such Bt toxins that end up in the soil and water from transgenic leaf litter may have negative impacts on soil and aquatic invertebrates and nutrient cycling processes, all aspects that deserve serious further inquiry.

A major environmental consequence resulting from the massive use of Bt toxin in cotton or other crop occupying a larger area of the agricultural landscape, is that neighboring farmers who grow crops other than cotton, but that share similar pest complexes, may end up with resistant insect.

Gene flow

Just as it occurs between traditionally improved crops and wild relatives, pollen mediates gene flow between GMCs and wild relatives or conspecifics despite all possible efforts to reduce it. Little is known about the long-term persistence of crop genes in wild populations or about the impact of fitness-related crop genes on the population dynamics of weedy relatives. The main concern with transgenes that confer significant biological advantages is that they may transform wild/weed plants into new or worse weeds. Hybridization of HRCs with populations of free living relatives will make these plants increasingly difficult to control, especially if they are already recognized as agricultural weeds and if they acquire resistance to widely used herbicides. For example:

— Transgenic resistance to glufosinate can be passed on from Brassica napus to populations of weedy Brassica napa, and persist under natural conditions.

— In Europe there is a major concern about the possibility of pollen transfer of herbicide tolerant genes from Brassica oilseeds to Brassica nigra and Sinapis arvensis.

Economic and Agronomic Implications

World-wide in 2000, transgenic herbicide resistant crops were planted on 74% of the 44.2 million hectares devoted to transgenic crops. In North America, transgenic glufosinate resistant cultivars of canola and corn, and transgenic glyphosate resistant cultivars of soybean, corn, cotton, and canola are now commercially available. Bromoxynil resistant transgenic cotton has also been developed. The so-called Round-up ready soybeans are the most prevalent GMCs. Transgenic herbicide resistance in crop plants simplifies chemically based weed management because it typically involves compounds that are active on a very broad spectrum of weed species.

Post-emergence application timing for these materials fits well with reduced or zero-tillage production methods, which can conserve soil and reduce fuel and tillage costs. However, HRCs also have significant problems.

— Reliance on HRCs perpetuates the weed resistance problems and species shifts that are common to conventional herbicide based approaches.

— Herbicide resistance becomes more of a problem as the number of herbicide modes of action to which weeds are exposed becomes fewer and fewer, a trend that HRCs may exacerbate due to market forces.

— Given industry pressures to increase herbicide sales, acreage treated with broad-spectrum herbicides will expand, exacerbating the resistance problem. For example, it has been projected that the acreage treated with glyphosate will increase to nearly 150 million acres. Although glyphosate is considered less prone to weed resistance, the increased use of the herbicide will result in weed resistance, even if more slowly, as it has been already documented with Australian populations of annual ryegrass, quackgrass, birdsfoot trefoil and Cirsium arvense.

— Perhaps the greatest problem of using HRCs to solve weed problems is that they steer efforts away from crop diversification and help to maintain cropping systems dominated by one or two annual species. Crop diversification can

a) reduce the need for herbicides
b) improve soil and water quality
c) minimize requirements for synthetic nitrogen fertilizer
d) regulate insect pest and pathogen populations
e) increase crop yields and reduce yield variance.

Thus, to the extent that transgenic HRCs inhibit the adoption of diversified cropping systems that include rotational crops, cover crops and green manure, they hinder the development of sustainable agriculture.

Ecological Risks of Bt Crops

Based on the fact that more than 500 species of pests have already evolved resistance to conventional insecticides, pests can also evolve resistance to Bt toxins present in transgenic crops. No one questions if Bt resistance will develop, the question is now how fast it will develop. Susceptibility to Bt toxins can therefore be viewed as a natural resource that could be quickly depleted by inappropriate use of Bt crops. However, cautiously restricted use of these crops should substantially delay the evolution of resistance. But is cautious use of Bt crops possible given commercial pressures that have resulted in a rapid rollout of Bt crops reaching 8.2 million hectares worldwide in 2000?

The refuge strategy of setting aside 20-30% of land to non-Bt crops to delay resistance is very difficult to implement regionally. Data from the Midwest shows that Bt corn saves on some insecticide use and yields are 2.4 Bu/acre higher than conventional corn but only under high European corn borer infestations. On the other hand organic corn growers use no insecticides and obtain yields (4.8-9 t/ha) similar or slightly higher than conventional farmers (5.0-7.1 t/ha).

Bacillus thuringiensis proteins are becoming ubiquitous, highly bioactive substances in agroecosystems. Most non-target herbivores colonizing Bt crops in the field ingest plant tissue containing Bt protein which they can pass on to their natural enemies in a more or less processed form. Polyphagous natural enemies (polyphagous: subsisting on many kinds of foods) that move between crop cultures are found to frequently encounter Bt-containing non-target herbivorous prey in more than one crop. This is a major ecological concern given previous studies that documented that Cry1 Ab adversely affected the predaceous lacewing Chrysoperla carnea reared on Bt corn-fed prey larvae. These findings are problematic for small farmers in developing countries who rely on insect pest control, which involves a complexity of predators in their mixed cropping systems.

Research shows that natural enemies can be affected directly through inter-trophic level effects of the toxin present in Bt crops. This raises serious concerns about the potential disruption of natural pest control, as polyphagous predators will encounter Bt-containing, non-target prey that move within and between crop cultivars throughout the crop season. Disrupted biocontrol mechanisms will likely result in increased crop losses due to pest's or to the increased use of pesticides by farmers with consequent health and environmental hazards.

The possibilities for soil biota to be exposed to transgenic products are very high. The little research conducted in this area has already demonstrated:

— There is long term persistence of insecticidal products (Bt and proteinase inhibitors) in soil.

— The insecticidal toxin produced by Bacillus thuringiensis subsp. kurskatki remain active in the soil, where it binds rapidly and tightly to clays and humic acids.

— The bound toxin retains its insecticidal properties and is protected against microbial degradation by being bound to soil particles, persisting in various soils for at least 234 days.

— The presence of the toxin in exudates from Bt corn and verified that it was active in an insecticidal bioassay using larvae of the tobacco hornworm.

Given the persistence and the possible presence of exudates, there is potential for prolonged exposure of the microbial and invertebrate community to such toxins, and therefore studies should evaluate the effects of transgenic plants on both microbial and invertebrate communities and the ecological processes they mediate. If transgenic crops substantially alter soil biota and affect processes such as soil organic matter decomposition and mineralization, this would be of serious concern to organic farmers and most poor farmers in the developing world.

These farmers cannot purchase or don't want to use expensive chemical fertilizers. They rely instead on local residues, organic matter and especially soil organisms for soil fertility (e.g., key invertebrate, fungal or bacterial species) which can be affected by the soil bound toxin. Soil fertility could be dramatically reduced if crop leachates inhibit the activity of the soil biota and slow down natural rates of decomposition and nutrient release.

Environmental Risks of Transgenic Crops

As long as transgenic crops follow closely the pesticide paradigm, such biotechnological products will do nothing but reinforce the pesticide treadmill in agroecosystems, thus legitimizing the concerns that many scientists have expressed regarding the possible environmental risks of genetically engineered organisms. The most serious ecological risks posed by the commercial-scale use of transgenic crops are:

— The spread of transgenic crops threatens crop genetic diversity by simplifying cropping systems and promoting genetic erosion;

— The potential transfer of genes from pesticide resistant crops (HRCs) to wild or semidomesticated relatives thus creating super weeds;

— HRC volunteers become weeds in subsequent crops;

— Vector-mediated horizontal gene transfer and recombination to create new pathogenic bacteria;

— Vector recombination to generate new virulent strains of virus, especially in transgenic plants engineered for viral resistance with viral genes;

— Insect pests will quickly develop resistance to crops with Bt toxin;

— Massive use of Bt toxin in crops can unleash potential negative interactions affecting ecological processes and non-target organisms.

The above impacts of agricultural biotechnology are herein evaluated in the context of agroecological goals aimed at making agriculture more socially just, economically viable and ecologically sound. Such evaluation is timely given that worldwide, there have been over 1,500 approvals for field testing transgenic crops (the private sector has accounted for 87% of all field tests since 1987), despite the fact that in most countries stringent procedures are not in place to deal with environmental problems that may develop when engineered plants are released into the environment. A main concern is that international pressures to gain markets and profits is resulting in companies releasing transgenic crops too fast, without proper consideration for the long-term impacts on people or the ecosystem.

Most innovations in agricultural biotechnology are profit driven rather than need driven, therefore the thrust of the genetic engineering industry is not really to solve agricultural problems, but to create profitability. This statement is supported by the fact that at least 27 corporations have initiated herbicide-tolerant plant research, including the world's eight largest pesticide companies Bayer, Ciba-Geigy, ICI, Rhone-Poulenc, Dow/Elanco, Monsanto, Hoescht and DuPont, and virtually all seed companies, many of which have been acquired by chemical companies. In the industrialized countries from 1986-1992, 57% of all field trials to test transgenic crops involved herbicide tolerance and 46% of applicants to the USDA for field testing were chemical companies.

Crops currently targeted for genetically engineered tolerance to one or more herbicides includes: alfalfa, canola, cotton, corn, oats, petunia, potato, rice, sorghum, soybean, sugarbeet, sugar cane, sunflower, tobacco, tomato, wheat and others. It is clear that by creating crops resistant to its herbicides a company can expand markets for its patented chemicals. The market for HRCs has been estimated at more than $500 million by the year 2000. Although some testing is being conducted by universities and advanced research organizations, the research agenda of such institutions is being increasingly influenced by the private sector in ways never seen in the past.

46% of biotechnology firms support biotechnology research at universities, while 33 of the 50 states have university-industry centers for the transfer of biotechnology. The challenge for such organizations will not only be to ensure that ecologically sound aspects of biotechnology are researched and developed (N fixing, drought tolerance, etc.), but to carefully monitor and control the provision of applied non-proprietary knowledge to the private sector so as to protect that such knowledge will continue in the public domain for the benefit of all society.

Super Weeds

Although there is some concern that transgenic crops themselves might become weeds, a major ecological risk is that large scale releases of transgenic crops may promote transfer of transgenes from crops to other plants, which may then become weeds. The biological process of concern here is introgression, that is, hybridization among distinct plant species. Evidence indicates that such genetic exchanges among wild, weed and crop plants already occur. The incidence of shattercane (Sorghum bicolor), a weedy relative of sorghum and the gene flows between maize and teosinte demonstrates the potential for crop relatives to become serious weeds. This is worrisome given that a number of US crops are grown in close proximity to sexually compatible wild relatives. There are also crops that are grown near wild/ weedy plants that are not close relatives but may have some degree of cross compatibility such as the crosses of Raphanus raphanistrum R. X Sativus (radish) and Johnson grass X Sorghum corn.

Total weed removal via the use of broad-spectrum herbicides may lead to undesirable ecological impacts, given that an acceptable level of weed diversity in and around crop fields has been documented to play important ecological roles such as enhancement of biological insect pest control, better soil cover reducing erosion, etc.. HRCs will most probably enhance continuous cropping by inhibiting the use of rotations and polycultures susceptible to the herbicides used with HRCs. Such impoverished, low plant diversity agroecosystems provide optimal conditions for unhampered growth of weeds, insects and diseases because many ecological niches are not filled by other organisms. Moreover, HRCs, through increased herbicide effectiveness, could further reduce plant diversity, favoring shifts in weed community composition and abundance, favoring competitive species that adapt to these broad-spectrum, post emergence treatments.

11

Impacts of Biotechnology on Biodiversity

Biodiversity today is the result of 3.5 billions years of evolution. All the living organisms we know today, as well as those that ever lived before, have developed from one original micro-organism through the processes of mutation and selection. Separate species arose when mutations between relatives no longer allowed for interbreeding, for instance after geographic separation. The vast majority of species that arose, probably more than 99%, disappeared again. In the long-term view there has been no such thing as sustainability, only change.

Today, with man's massive influence on the globe, change and the loss of biodiversity is much faster than at any time before, making concern with sustainability important. Biodiversity can be distinguished at three different levels, ecosystems, species and genes. But there is no generally accepted definition of biodiversity and there is no general consensus on how to appreciate changes in biodiversity from scientific, political and/or normative perspectives. The number of species of animals, plants and micro-organisms today is probably 10 million or more, of which only 1.4 million have yet been named.

Virtually all the 40,000 vertebrate animals and most of the 250,000 higher plant species are known. On the other hand there are likely to be over a million species each of fungi and nematodes (thread-like worms), and several million insects, of which only 70,000, 13,000 and 950,000 respectively have been identified. Only about 5,000 bacteria and viruses have been identified individually, yet their total number may also be well in excess of one million.

The various species of plants and animals do not live an independent existence but are associated in specific communities and ecosystems to form more or less stable associations. One such association, for instance, is the tropical rain forest which is generally thought to have the highest degree of biodiversity. Biodiversity needs to be considered both in terms of the number of species present and also of how many individuals there are of each of the species.

Often the number of species in a given ecosystem is taken as a measure of the biodiversity of that system because other criteria are more difficult to apply. In addition, it is still not clear whether the total amount of living matter at a given location, the biomass, is generally dependent on biodiversity: some researchers claim that loss of species is not compensated by additional growth of other species and thereby leads to a reduction in total biomass. In agriculture, some 7,000 species of plants are used by farmers throughout the world, but only 30 species provide 90% of our calorific intake, the top three crops being wheat, rice and maize (corn).

Within the predominant crop species there are many hundreds of thousands of varieties (landraces) adapted to local climates and farming practices. Much of this large crop diversity is important for providing starting material for breeding. However, the genetic diversity in crops is much less broad than that of plants or animals living in the wild which indicates the importance of wild species for agricultural breeding programmes. Human population growth. The world population has risen from 2.5 billions in 1950 to 6 billion today and is expected to reach some 9 billion in 2050. 800 million people are currently undernourished.

Over 95% of the expected population increase will be in the less developed countries (LDCs) and most will occur in their cities, requiring more living space, water, energy, wood, food and services. For 1995 to 2020 the largest relative population increase (80%) is expected in sub-Saharan Africa rising from 500 to 900 million. Whether the HIV/AIDS epidemic in this region will substantially affect population dynamics is not clear as most of those currently infected are in the reproductive age range.

RISKS TO NATURAL BIODIVERSITY

There are many genetic transformations in crops, such as altered starch, oil, and fat content, which will probably have little or no adverse impact on biodiversity. Most of the present generation of GM crops carry transformations for the insertion of genes for herbicide tolerance and insect resistance into existing crop varieties. Genetically modifed herbicide tolerant (GMHT) and genetically modifed insect resistant (GMIR) crops are being used commercially now over some 40 million hectares worldwide.

GENE FLOW AND TRANSFER OF TRAITS

Recent research confirms that genes introduced into some genetically improved crops will spread into related native species. Gene transfer is almost inevitable from crops that have interfertile relatives in adjacent natural ecosystems, but not from crops such as the maize and cereals grown in Europe, whose closest relatives are on the other side of the oceans. After all, genes have been moving for many years from "conventionally" bred crops to wild relatives; for example, in the UK hybrids occasionally occur between oilseed rape (Brassica napus) and native species like wild turnip (B. rapa).

The difference of course is that genes inserted into GM crops are often derived from other phyla, giving traits that have not been present in wild plant populations, and if introduced accidentally, may change the fitness and population dynamics of hybrids between native plants and crops, eventually backcrossing into the native species and becoming established. So the issue is not so much the rate of gene flow, rather the impact that this might have on agriculture and biodiversity. Conventional plant breeding, using mutagenesis and embryo rescue techniques, also produces lots of completely new genes in crops, about which we know very little.

Interestingly, these are often the very crops being used by organic farmers and being sold as "natural foods"! Most geneticists would argue that most "foreign" genes introduced into crop/native hybrids would in fact decrease their fitness in the wild, leading to rapid selection of these genes out of the population. This is particularly true of genes designed to prevent germination of saved seed, like the so-called terminator gene—if this were to "escape" it would commit instant suicide and certainly not spread into the natural world as has been suggested by some anti-GM campaigners.

There is no difference to the farmer between buying seed with terminator technology and buying hybrid seed, because neither can be saved and grown next year. There is a serious issue about whether farmers in the developing countries should become locked into a cycle of dependence on patented seed, but the genetics of this technology is not a direct environmental threat. Transfer of certain genes, such as resistance to insects, fungi and viruses could increase fitness (ability to reproduce) of any resulting hybrids, possibly forming aggressive weeds or plants that swamp wild populations.

Weeds having tolerance to a range of herbicides could also emerge; these would be difficult to control in agriculture, or in natural ecosystems like grasslands. Farmers may eventually need mixtures of herbicides to control them, causing yet more damage to biodiversity. There is already evidence from North America that this "multiple tolerance" and resistance to herbicides is beginning to emerge. If nontarget plants acquired insect resistance from GM crops, they could damage food chains dependent on insects feeding on previously nontoxic wild plants.

Not only would there be a direct effect, for many insects are entirely dependent on single plant species, but acquisition of resistance in wild plants may change their population dynamics, increasing the risks of them invading agricultural land and natural ecosystems.

These ecological genetics principles also apply to virus and fungus resistances. This is an even more serious issue for developing countries where control of invasive plants is a major problem for subsistence farmers and may have implications for biotopes of global importance. The science we urgently need to be able to assess these risks is simply not being done. At the moment we do not know what effect escaped genes might have on natural and farmland ecosystems. This lack of science is disturbing, given the commercial pressure and rapid timetable for the introduction of GM crops into our landscapes.

Science will never tell us everything about what might happen, but no science will tell us nothing. Genetic transfer to native ecosystems not only carries ecological risk, but also undermines fundamental reasons for conserving plants and their dependent ecosystems in situ. Our understanding of ecological genetics depends on research on gene pools of species making up native ecosystems, and the genetic code of each wild species holds information which may eventually benefit us. So-called "genetic pollution" of native gene pools raises some legitimate questions about the loss of basic scientific resources.

As scientists, we are keenly interested in the genetics of native populations, so to add genes from other phyla unwittingly and randomly to gene pools is not necessarily a good idea. There is clearly a need to set up effective monitoring systems to detect gene transfer and research to assess ecological impacts. Research in this area would be in the interests of both the industry and the environment. It would be far better for biotechnology companies to produce the next generation of GM crop plants with in-built mechanisms, such as pollen incompatibility, to prevent gene flow.

Perhaps the ecologically simplest way to ensure genetic isolation is to make sure that wherever possible plants used for genetic modification are unrelated to native species and edible crops whose center of origin is within the intended market territory. Biotechnology companies should start thinking now about which plants are chosen as platforms for biomedical and industrial product transformations. If biotechnology is ever to become a standard technique for plant breeding.

Genetically Improved Crops

The prospect of gene transfer causes concern for crops that have wild relatives in the same ecosystem, and occupies reams of headline comment in the press. Perhaps of greater importance is the fact that management of some genetically improved crops would be very different from conventional intensive agriculture or organic farming. In the United States, genetically modifed herbicide tolerant (GMHT) crops are grown under a regime of broad-spectrum herbicides applied during the growing season. Farmers report almost total weed elimination from GMHT crops, which include cotton, soybean, maize, beet, and oilseed rape. They also report substantial reduction in herbicide use.

Recent research in the United Kingdom confirms that weed control in GM beets and other GMHT crops is likely to become much more efficient. These results are hardly surprising since this is the main purpose behind the technology. This GMHT system will soon be available, at least experimentally, for virtually all mainstream agricultural crops, including vegetables. Broadspectrum herbicides used on commercial scale GMHT crops during the growing season may be far more damaging to farmland ecosystems than the selective herbicides they might replace. Using these herbicides in the growing season may also increase the impact of spray drift onto marginal habitats such as ancient hedgerows (field margins common in Europe) and watercourses. It is not only the volume of herbicides that is the issue but their efficiency and impact on wildlife.

When insect resistance and herbicide tolerance are combined in the same crop variety, there may be few insects capable of feeding on the crops and few invertebrates and birds would be able to exploit the weed-free fields. In Europe we already have massive declines in farmland birds, with several previously common species now close to extinction. The problem with assessing the environmental impact of these changes in management is that the regulatory system and the public has very little scientific data on which to assess the real risks, and potential benefits, from adopting GMHT crop systems.

Formal risk assessments submitted by the biotechnology companies as part of the regulatory process deal with this issue inadequately. In the United Kingdom, the Department of Environment, Transport and the Regions and the Ministry of Agriculture, Food and Fisheries have realised this, changed the regulatory system, and commendably have started some field-scale experiments to try to answer some of these important questions. The development of new crops with improved tolerance to abiotic factors (such as drought, salinity, frost) and the potential advent of 'pharmed' crops producing vaccines and GM biomass systems, may also change crop management, perhaps increasing demand for arable land in the long term, and putting further pressure on natural biodiversity on marginal land.

Agricultural Intensification

If we want to make predictions about how intensification enabled by GM crops could affect biodiversity, we can turn to evidence of declines in farmland plants, insects, and birds resulting from agricultural intensification in Europe over the past 30 years. Factors responsible include abandoning traditional crop rotations, increased pesticide efficiency and drift, use of artificial fertilizer, drainage, and intensification of soil cultivation. There is overwhelming evidence demonstrating that the use of more effective pesticides (including herbicides) over the past 20 years has been a major factor causing serious declines in farmland birds, arable wild plants, and insects.

Pesticides not only have direct toxic effects on wildlife but they also enable modern crop management changes to take place. Winter-sown crops, for example, rely heavily on effective fungicides. Thirty years ago winter sowing was unknown in the United Kingdom and winter stubbles were widespread, providing an essential food source for wintering flocks of birds. There are many examples of declines in farmland wildlife in the UK and these are typical of intensively managed farmland throughout Europe.

It is important to remember that although these declines in biodiversity have been severe in many intensively managed areas, there are still viable populations of many farmland-dependent species throughout Europe. Some of these, however, are only just surviving the impact of intensive agriculture. Twenty-five of the 200 species of British "arable plants" are now "Nationally Scarce" and a further 24 are "of conservation concern" and included in the 1983 IUCN Red Data Book (RDB). Not only have many arable plants become threatened but there has also been a marked shift towards a less diverse, grass-dominated flora.

More effective herbicides are responsible; similar trends have been observed elsewhere in Europe. Changes in herbicide practice have also been a major factor in reducing the distribution of insects such as the common blue butterfly, the larvae of which feed on broad-leaved weeds. Over half of British farmland birds are now in serious decline and 13 are red-listed. The 78 percent drop in grey partridge (Perdix perdix) numbers observed in the United Kingdom between 1972 and 1996, has been directly attributed to increased herbicide and pesticide efficiency. Skylark (Alauda arvensis) populations have declined by 75 percent over this period mainly due to increased pesticide efficiency. Recent research implicates agricultural intensification in the decline of other songbirds.

Besides the aesthetic and scientific reasons for conserving biodiversity within and around agricultural crops, there is another important utilitarian reason for wanting to do so. This is, the need to maintain the food chain links between native species and crop systems. This link is vital if we are to preserve the function of biodiversity to deliver early warning of dangers in crops or the chemicals used to manage them.

Without these links, we are unlikely to be able to detect any dangers arising from the new agriculture by monitoring wildlife; the first organism in the food chain will increasingly be Homo sapiens. This "natural early warning system" has served agriculture and the public very well over the past 50 years. It detected the toxicity of DDT and aldrinbased organochlorine pesticides and showed up the potentially lethal effects of PCBs before toxic levels built up in humans. This is not just an issue for the industrial countries. It is a natural alarm system which is probably the most cost-effective way of monitoring environmental safety in developing countries.

Loss of Biodiversity

Loss of biodiversity can be measured by a loss of individual species, groups of species or by decreases in numbers of organisms. The major threats to global biodiversity are, firstly, habitat loss (mostly through expansion of cultivated land and of cities and roads) and, secondly, the introduction of exotic species. Habitats can also be damaged by flooding, lack of water, climate changes etc., natural or man-made. Since tropical humid forests are particularly rich in biodiversity, their destruction is disproportionately damaging. It is estimated that only half of the original 16 million km^2 of these forests a century ago are left, with about one million km^2 being destroyed every 5 to 10 years.

Biodiversity is not homogeneously distributed but rather there are areas with particularly plentiful biodiversity ("hotspots") of particular interest for conservation. Imported plant and animal species threaten the native ones by being highly competitive and often by lacking local predators. One of the most extreme examples is seen in the pampas of Argentine, a flat grassland with a moderate climate, from which nearly all the native grasses have disappeared and have been replaced by European plants. This invasion was brought about by European farmers, bringing animals and crops, as well as accidentally spreading many different weeds and was already noted in 1833 by Charles Darwin.

Biological control agents are often introduced into agricultural ecosystems intentionally to control pests or weeds without resorting to chemical controls. Whilst there have been many welcome successes, such systems may also go wrong. One example is the introduction of the seven-spot ladybird which was intended to control the Russian wheat aphid in the US. This proved to be a competitor of the native ladybird, which then disappeared. The mongoose, an Indian mammal, was introduced to several islands (Fiji, Mauritius, Hawaii) to keep rats and snakes under control: it lead to the extinction of several endemic birds, reptiles and amphibians.

Introduced wasps have been seen to kill and possibly extinguish endemic butterflies. Are transgenic plants as such prone to spread? In the longest term experiment so far, four different crops (oilseed rape, potato, maize and sugar beet) were grown in 12 different habitats and monitored over a period of 10 years. In no single case were the genetically modified plants found to be more invasive or more persistent than their conventional counterparts. However, one would not expect this to be the case unless the transgene increased its fitness in the wild.

There is no plausible reason why crops that have for centuries depended for survival on human care should become weeds because of the addition of one or a small number of well-characterised genes, in addition to the many thousands of genes they already carry. Even so, monitoring of transgenic crops needs to continue for more than ten years. Large acreages of a single variety without rotation, should be avoided with any crop, since monocultures are more prone to disease and pest outbreaks.

Conservation

Conservation may be in situ in a more or less natural or habitat or ex situ in some purposebuilt environment depending on the particular case. In situ conservation involves the maintenance and protection of natural habitats while botanical gardens, seed banks and zoos are used for ex situ conservation. Conserving a substantial fraction of the tropical rain forests would still allow half or so of their indigenous species to be preserved by selection of the most appropriate "hot spots".

Protecting large tracts of land poses major socio-economic and political problems. How forests can be kept free from encroachment by hungry people in search of potential farmland is far from clear. A viable strategy may be to find sustainable livelihoods for rural populations compatible with conserving tropical rain forests and this is being attempted with mixed success. Policing alone will not be successful over vast territories, as seen in the combating of drugs. Conservation also embraces agricultural biodiversity such as crop varieties, land races (local varieties), semi-domesticated varieties and crop relatives. The role of indigenous communities in maintaining agrobiodiversity is stressed by the Global Biodiversity Assessment and the Leipzig Plan of Action, two recently concluded international agreements.

The Convention on Biological Diversity (CBD)

The United Nations adopted the Convention in 1992 and it entered into force in 1993. This is the first time that a large majority of States, though not the USA, have agreed to a legally binding instrument for biodiversity conservation and the sustainable use of biological resources. A radical change brought about by the CBD is the recognition that States have a sovereign right over biodiversity within their own territory, while previously living organisms were considered the common heritage of mankind.

Under the terms of the CBD, living organisms or their products may only be removed from a country under mutually agreed conditions. Action is delegated to the national level obliging States to assess biodiversity, enact legislation for its conservation in situ and ex situ, and to enforce legislation within national boundaries. Biotechnology is particularly affected by Articles 16 and 19 of the CBD since they require a fair and equitable sharing of benefits derived from the use of genetic resources. This includes providing facilities and financial means for technology transfer and open access to scientific and technical information.

The CBD has only been in operation for a few years and so it is too early to assess its long-term effects. As far as biotechnology and "bioprospecting" are concerned, it will take time to establish smooth administrative procedures to allow simple routine implementation of close collaborations. Only if national authorities from countries rich in biodiversity as well as pharmaceutical and other companies see the advantages to be gained from such collaboration will the system expand. One example of a joint effort in "bioprospecting" is the search for specific active ingredients of plants by the Merck Company in the tropical forests of Costa Rica which brought the country 2 million USA dollars over a five year period as well as the potential of royalties if profitable products emerge.

The Cartagena Protocol on Biosafety

The CBD provided a basis for developing an international agreement regulating the crossboundary movement of living GMOs, called "living modified organisms" in the Protocol, which after a lengthy debate was agreed on in Montreal in early 2000. The main aim of the Protocol is to ensure that GMOs would not endanger biodiversity in the recipient countries; in addition it also covers potential threats to human health. In practice the Protocol will primarily affect the importation of transgenic seeds. It requires the exporting country to provide enough scientific information to allow the importing country to judge risks and then to issue, if the product is judged safe, an "Advance Informed Agreement".

Based on the "Precautionary Principle" import licences can be refused, even when there is no clear-cut proof that the product is dangerous to human health or the environment, but such decisions need to be based on a risk assessment. The Protocol is construed to be as important as the World Trade Organisation (WTO) agreements which do not allow countries to prevent the import of products unless there are sound scientific reasons showing that the product may cause harm to human health. This apparent lack of priority between the CBD and WTO agreements may lead to some future difficulties. The Protocol will become operational when it is ratified by 50 nations, presumably in 2002.

Biotechnology can be used as a tool for acquiring scientific knowledge or to intervene directly in plant and animal breeding and, in particular, to transfer genetic information from any other organism to a crop or farm animal. Taxonomy, one of the sciences necessary to study biodiversity, uses molecular markers to identify individual strains of organisms or to identify species in much the same way as is done in forensic medicine to identify criminals. In seed banks genetic fingerprints are used to establish the origin of a seed or the relatedness of plant varieties. Biotechnology is also useful for following genetic markers in plant and animal breeding done by conventional means.

By analysing a few cells of the newly born calf or of the newly sprouted crop and looking for the presence or absence of certain genes it is possible to predict properties of the progeny which will show up only later in life such as the characteristics of a cow's milk or a crop's expected resistance to a plant disease. These applications of biotechnology to farm animals (not to humans) are hardly controversial.

Native Biodiversity and Biotechnology

Biodiversity in the wild has been massively reduced in the industrialised countries over many centuries and about half the tropical rain forests have already been destroyed. How then can the rest be preserved, given that the world population is increasing rapidly and more food has to be produced regionally? Yields of cereals in LDCs have gone up very considerably in the last forty years primarily as the result of the Green Revolution.

However, the annual growth increases in cereal yields in LDCs have slowed down from about 3% during 1967-82 to about 1% per year from 1993 onwards. This lowering of the rate at which yields have increased means that productivity will probably not keep up with the demands of increased populations. The consequences for biodiversity are devastating as more land will be required for farming which will primarily come from areas with high native biodiversity, and in particular tropical rain forests.

The single most promising way to avoid habitat destruction is to increase farm yields in a process that has been called the "Second Green Revolution". Several components will be required including training and education of farmers (in particular of women who do most of the farm work in LDCs), more favourable economic and political climates, availability of farm credit schemes, etc. In addition, technical contributions will be necessary and, in particular, improved seed produced either by traditional crop breeding or by modern biotechnology.

Reliance will have to be more on the latter since traditional breeding appears to have reached a plateau in yield and is slower, less precise and only feasible when interbreeding is possible. So agricultural biotechnology, which is viewed frequently in the public debate as harmful to biodiversity, is, paradoxically, likely to contribute to conserving it. A more limited concern that largely affects Northern Europe is the conservation of native plants and animals, in particular birds, in farmed areas. Their habitats are fields, hedges, roadsides and fallow land where they depend for food on insects and seeds produced by weeds in or near crops.

Computer models suggest that more intense weed control measures may lead to smaller amounts of seeds being available to birds. This depends far more on weed management regimes rather than on transgenic plants. Herbicide tolerant beets allow farmers to control weeds later by treating after the seedlings have emerged. The more efficient methods may allow setting aside of more land. Setting aside more farmland requires financial incentives.

Gene transfer between cultivated and wild plants has always occurred within the limits of species if the two types of plants were in close proximity and flowered at the same time. No new problems can be expected from transgenic plants except if the gene transferred from the GMO to the wild plant significantly increased the biological fitness of the recipient. This generally seems unlikely a priori and can, and should, be checked experimentally by risk assessment. Some people feel there is a crucial difference between transfer of genes from conventionally bred plants to their wild relatives and transfer of man-made transgene constructs from a GM plant into a gene pool.

Whether ecological risks may result from the introduction of a man-made gene construct into the gene pool of a species is a different question. Farmers all over the world use a remarkable number of landraces and varieties of many different crops. These are usually closely adapted to the local climate and topography, and are also used to produce foods for different purposes. Will the traditional varieties and landraces disappear when, and if, transgenic crops are introduced? Will only one or very few varieties dominate in the field? To judge by what has happened on the US soybean market, this will not be the case.

Although virtually all herbicide resistant soybeans derive from a single transformation event in Monsanto's Roundup Ready soya, hundreds of different varieties have been derived from it by different seed companies using traditional breeding to develop soya varieties that suit different climatic and soil regions. This shows that at least with this dominant transgenic crop biotechnology has not led to a loss of agricultural biodiversity.

On the other hand, judging from the last 100 years in Europe, the diversity of varieties of farm animals and crops has, in fact, diminished considerably, but not because of any biological hazard emanating from some of the breeds, but because farmers need to produce economically. Many old varieties of apples have disappeared, for example, because of the preference of food retailers and consumers. The rapid consolidation in the global seed market may be a concern in this respect and anti-trust legislation may be needed to prevent the dominance of oligopolies with limited competition between few producers. It can be argued that only wealthy farmers will be able to afford transgenic seeds.

However, the technology is well suited to small farmers, since it is packaged in a small seed grain, which can be reproduced locally. However, if the seeds are to reach small farmers, there need to be many seed outlets (public cooperatives or private companies) and adequate agricultural extension programmes.

Economic and Political Conditions

The original green revolution which improved wheat and rice production in Asia was initiated by research carried out in the public sector. Today's innovative research in biotechnology is mainly done by a few large companies with modest contributions from public sector and small companies. Companies strive for intellectual property rights (IPR) through patenting for return on their research investment. This is largely unavoidable with the present IPR systems.

Provisions need to be made so that the agricultural research institutions of LDCs obtain access to information as well as patented materials and procedures needed for regional farmers. Unfortunately financial support to the agricultural research systems of LDCs has been substantially cut in the last ten years: this trend endangers world food security and needs to be reversed.

A limited number of transfers of specific patents to LDCs has already occurred from companies such as Monsanto and Syngenta for producing virus-resistant sweet potatoes and Vitamin A enriched golden rice. This approach, including the possibility of enforced licensing, is probably more practicable than changing the world's patenting system which does, however, need to be adapted in order to deal with living organisms. A new issue arising in the field of IPR is the attempt to protect traditional, indigenous knowledge through novel international legislation.

Lack of food appears to be, and is largely, a problem of distribution; there is not sufficient food available to those who need it. The poor simply do not have the money to buy enough. Since in LDCs most of the poor still live on the land, it is here that production has to be increased so that there is not only enough for survival but a surplus which can be sold to buy other goods and services. Increasing local agricultural productivity is the most effective way to do this and should, according to many economists of the public sector including those of the UNDP, embrace modern biotechnology as an important component.

12

Effects of Herbicide Tolerant Crops

Modern agricultural practices have been broadly linked to declines in biodiversity in agro-ecosystems. This has been found to be true for a wide variety of taxonomic groups, geographic regions and spatial scales. More specifically, various researchers have found significant correlations between reductions in biodiversity at various taxonomic levels and agricultural intensification.

Effects of agricultural intensification undoubtedly highlight many contributing factors, which are addressed individually in the following sections, and include the cropping pattern, the frequency of tillage, the amount and nature of fertilizers used and the amount and nature of pesticides applied (particularly insecticides and herbicides). However, it should be noted that all of these factors are interrelated to a greater or lesser degree, often causing negative synergies. There is no doubt that many human, social and cultural factors must be taken into account, but nevertheless, in all cultures it is uncontested that habitat conversion is acceptable to provide more food and settlement for our own needs.

The kinds of potential impacts of GM crops fall into the classes familiar from the cultivation of non-GM crops (invasiveness, weediness, toxicity or biodiversity). It is likely, however, that the novelty of some of the products of GM crop improvement will present new challenges and perhaps opportunities to manage particular crops in creative ways'. Impact of agricultural biotechnology on biodiversity With the introduction of GM crops, concern has been raised that overall genetic diversity within crop species will decrease because breeding programmes will concentrate on a smaller number of high value cultivars.

The introduction of transgenic cultivars in agriculture has not significantly affected levels of genetic diversity within crop species. The genetic structure of the elite soybean population in North America, using coefficient of parentage (CP) analysis. The introduction of herbicide-tolerant cultivars with the Roundup Ready® trait was shown to have had little effect on soybean genetic diversity because of the widespread use of the trait in many breeding programmes. Only 1% of the variation in CP among lines was related to differences between conventional and herbicide-tolerant lines, whereas 19% of the variation among northern lines and 14% of the variation among southern lines was related to differences among the lines from different companies and breeding programmes.

Genetic uniformity among cotton varieties in the USA, there genetic uniformity had not changed significantly with the introduction of transgenic cotton cultivars. Genetic uniformity actually decreased by 28% over the period of introduction of transgenic cultivars. GM crops should be held responsible for a biodiversity decline within crops, are not very convincing. It remains to be said that the continued use of locally adapted traits gained in traditional breeding should have a more important role than it does at present.

Biotechnology represents a tool for enhancing genetic diversity in crop species through the introduction of novel genes. This does not aim at the single transgene inserted, but is based on the fact that beneficial characters can now be inserted in a variety of crops that have been neglected because of the limitations of traditional breeding methods, which failed to enhance the traits. There is great potential to achieve drought tolerance in vegetables and to avoid post-harvest losses in African grain staple crops.

Conservation Tillage

The soil in a given geographical area has had an important role in determining agricultural practices since the time of the origin of agriculture in the Fertile Crescent of the Middle East. Soil is a precious and finite resource. Soil composition, texture, nutrient levels, acidity, alkalinity and salinity are all determinants of productivity. Agricultural practices can lead to soil degradation and the loss of the ability of a soil to produce crops. Examples of soil degradation include erosion, salinisation, nutrient loss and biological deterioration. It has been estimated that 67% of the world's agricultural soils have been degraded. It is also worth noting that soil fertility is a renewable resource and soil fertility can often be restored by several years of careful crop management.

In many parts of the developed and the developing world tillage of soil is still an essential tool for the control of weeds. Unfortunately, tillage practices can lead to soil degradation by causing erosion, reducing soil quality and harming biological diversity. Tillage systems can be classified according to how much crop residue is left on the soil surface. Conservation tillage is defined as any tillage and planting system that covers >30% of the soil surface with crop residue, after planting, to reduce soil erosion by water.

The value of reducing tillage has been recognised for some time but the level of weed control a farmer required was viewed as a deterrent for adopting conservation tillage. Once effective herbicides were introduced in the latter half of the 20th century, farmers were able to reduce their dependence on tillage. The development of crop varieties tolerant to herbicides has provided new tools and practices for controlling weeds and has accelerated the adoption of conservation tillage practices and 'no-till'

at the country elevator when loading and to test every ship sublot when loading at the export elevator. This testing strategy results in average rejection rates at the importer of 1.75% with less than .02% of lots containing adventitious commingling remaining in importer flows after testing at the importer (due largely to test accuracy).

The proportion of flows in the Non-GM channel declined from .80 at the farm level to an average of .696 at the importer. Thus, on average, over 10% of Non-GM shipments are diverted to the GM segregation throughout the handling system.

This illustrates the risk of incorrectly rejecting shipments throughout the system. Most of this occurs after unloading at the export elevator. Further, there are 95% confident that diversions of Non-GM to GM shipments should range from about 8% to 17% of shipments. This diversion takes place due to large samples containing units with both adventitious commingling and Non-GM which are represented by a single test, adventitious commingling which occurs in the system, and through effects due to test accuracy.

Additional system costs for testing and discounts for rejection at the importer in the base case were 1.4 c/bu. The approximate components of these costs are: testing of every 5th railcar loaded at the country elevator, .037 c/bu; testing of every ship hold at the export elevator, .282 c/bu; testing of every ship hold at the importer, .259 c/bu; and rejection cost at the importer of .808 c/bu.

If this cost was absorbed solely by the Non-GM bushels, the costs average 2.0 c/bu. The cost of the system includes both additional system costs and the risk premium. Adding these two cost elements results in total costs of 2.4 c/bu when measured across all bushels and 3.4 c/bu when attributed solely to Non-GM bushels. These costs only reflect additional costs of testing and rejection within a system of Non-GM/GM wheat. Other costs could include costs for additional segregation, monitoring, etc., but were not included here.

In the base case, mechanisms are used to elicit information from growers on the GM content of their grains. In fact, this function would normally be included in "closed loop" marketing plans. This facilitates segregation at the point of first receipt, albeit at an allowed risk of adventitious commingling at the grower level and due to grower truth-telling. If such a mechanism were not developed, initial handlers would have greater uncertainty upon receipt, which in turn could impact the level of adventitious commingling due to the inability to segregate GM from Non-GM without testing. To simulate this impact, there developed a model without variety declaration which included a higher rate of adventitious commingling (20%, which is equal to the percent of GM adoption) at the point of first receipt.

With no variety declaration, the optimal testing strategy included testing of every 5th unit at the country elevator, both when receiving grain from growers and when loading railcars; and testing every 5th railcar when received at the export elevator and every hold when loaded at the export elevator. Rejection rates at the importer increased from 1.75% for the base case to 2.34% with no variety declaration.

The largest impact was on the proportion of Non-GM in the system. When flows reach the importer, only 30.6% of flows were Non-GM. Thus, a system with no variety declaration results in significant diversion of flows from Non-GM to the GM segregation. This occurs throughout the system, but is concentrated at the country elevator level. These are reflected in the costs when attributed to Non-GM bushels. Costs of testing and rejection for Non-GM bushels increased from 1.99 c/bu for the base case to 4.38 c/bu with no variety declaration. Total costs for Non-GM bushels similarly increased from 3.36 c/bu for the base case to 5.70 c/bu with no variety declaration.

A case was also developed where no testing and no variety declaration was allowed. This was used to reflect the risks inherent in the system and the value of testing. With no testing allowed, rejection rates at the importer were 10.10%. This is significantly higher than either the no variety declaration case or the base case. Total costs per Non-GM bushel were also significantly higher than either of the other cases.

Effect of Price Differentials

In corn and soybeans discounts have evolved to be about 10% of the value of grain. In the base case, discounts were applied which represent 10% of the value of wheat and added logistical costs to go to an alternative market. However, these are determined in part by contract specifications of individual buyers and by cumulative interaction of all buyers, sellers, and impacts of technology costs. Two cases were developed, one with lower penalties (0-10 c/bu) and a second with higher penalties (100-150 c/bu). A third case was developed to examine effects of additional discounts applied at loading locations (country and export elevators) when lots are identified as GM and diverted. Lower penalties resulted in an optimal testing strategy which was less intensive.

Optimal testing occurred at the same locations (both country elevator and export elevator loading); however, sampling at the export elevator was less intensive (every 5th unit versus every unit in the base case). This less intensive testing is reflected in a higher rejection rate, which increased from 1.75% in the base case to 7.87% with lower discounts. Higher penalties resulted in the same optimal testing strategy as the base case.

The third case applied additional penalties at loading and resulted in a more intensive testing strategy. Tests were conducted at the same locations; however, every unit was tested at both the country and export elevator when loading. The effect of this more intensive testing strategy for this case resulted in lower rejection rates at the importer (1.68% versus 1.77% in the base case).

Total costs (Additional + Risk Premium) when attributed to the Non-GM bushels increased as the level of the penalty for rejection increased. With the higher penalty (100 to 150 c/bu), total costs per Non-GM bushel were 5.19 c/bu while the low penalty rate had total costs of only 1.6 c/bu.

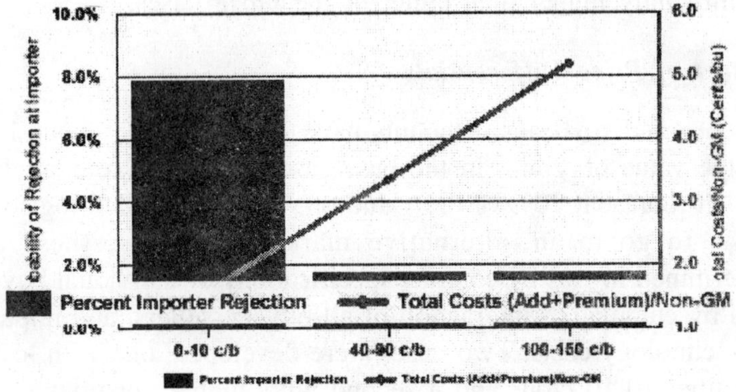

Figure 2. Effects of Penalty for Rejection on Rejection Rates and Total Costs per Non-GM Bushel.

This indicates that if the penalty for being out of specification for GM is minimal, the optimal response of decision makers is to test less often and accept higher rejection rates. As the penalty increases, decision makers respond with strategies which include greater testing intensities and lower rejection rates. Costs would be higher per Non-GM bushel and require higher risk premiums.

GM Adoption Rate

Adoption for GM in the base case was assumed to be 20% and was parameterised based on expected buyer aversion. However, this will vary with market forces, and most important, will vary geographically, both of which would impact the optimal testing strategy. Thus, the range of GM adoption was examined for cases with 10% and 50% GM adoption and compared to the base case.

Optimal testing strategies for 10% GM adoption were the same as for the base case. However, with GM adoption of 50%, the optimal testing strategy included testing at the country elevator and export elevator when loading and added additional testing (every 5th unit) when grain is received at the country elevator. As the level of GM adoption increased, the proportion of samples rejected at the importer also increased from a low of 0.17% for the 10% GM adoption to 2.12% with 50% GM adoption.

Costs of testing and rejection, when estimated over all bushels, declined as the level of GM adoption, increased. For GM adoption of 10%, additional costs attributed to all bushels were 1.53 c/bu and declined to 0.97 c/bu with 50% GM adoption. When costs are attributed only to Non-GM bushels, the additional testing and rejection costs increased from 1.94 c/bu for 10% GM adoption to 2.56 c/bu for 50% GM adoption. This occurs largely due to the lowering of the proportion of samples that are tested as the rate of GM adoption increases.

Farmers are assumed to declare GM content at the point of delivery. This allows the first handler to segregate and would be typically governed by some type of contractual relations and/or elevator imposed mechanism. In the base case, farmers were truthful in their declaration 95% of the time.

This was represented in the model by a triangular distribution with a minimum value of 80%, most likely 95% and maximum of 100%. Two cases were developed to examine the effect of reductions in farmer truth-telling. One case has truth-telling represented by a triangular distribution with minimum of 40%, most likely value of 50% and maximum of 60%, while the second case has a minimum of 65%, most likely value of 75% and maximum of 85%.

As farmer truth-telling declines, optimal testing strategies results in increased testing. Both cases with lower truth-telling included testing of every 5th unit at the country elevator when receiving in addition to testing at the country and export elevators when loading.

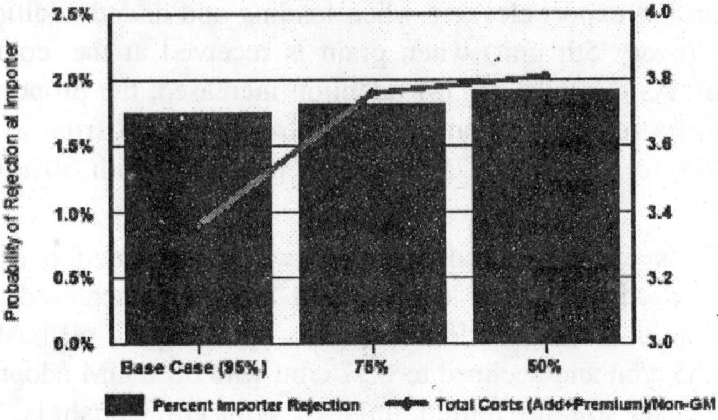

Figure 3. Effect of Grower Truth-telling on Rejection Rates at Importer and Total Costs per Non-GM Bushel.

Rejection rates at the importer increased from 1.75% in the base case to 1.91% for the lower truth-telling case (Figure 3). Also, the proportion of flows at the importer that were Non-GM declined from 69.6% in the base case to 58.0% with the lower truth-telling case. Thus, there is greater false rejections in the system as grower truth-telling decreases. Total costs for Non-GM bushels also increased as farmer truth-telling declined. Total costs for the lowest farmer truth telling were 3.81 c/bu for Non-GM bushels versus 3.36 c/bu for the base case.

In the base case the type of test was assumed to be a strip test at the country elevator and Polymerise Chain Reaction (PCR) tests at the export elevator and for importers. This assumption was relaxed and a case developed which in addition to choosing where and how often to test, the model also chooses what test to apply based essentially on the cost and risk. In this case, the choice of test was limited to choice at the country and export elevators. The optimal testing strategy when there was a choice of test included testing every 5th unit at the country elevator when loading, every 5th unit at the export elevator when receiving, and every unit at the export elevator when loading.

At all locations the strip test was chosen, which has a lower cost and test accuracy than the Polymerise Chain Reaction (PCR) test. This is reflected in a lower proportion of flows in the Non-GM segregation at the importer (65.4%) than in the base case (69.6%) and a higher rejection rate at the importer (2.08% versus 1.75% in the base case). Costs on all measures declined when the choice of test type was allowed. Testing and rejection costs across all bushels declined from 1.39 c/bu in the base case to 1.23 c/bu when choice of test was allowed.

The risk premium required for decision makers to be indifferent between a dual handling system and a Non-GM system declined from 0.96 c/bu in the base case to 0.80 c/bu when the choice of test is allowed. Total costs for Non-GM bushels declined from 3.36 c/bu in the base case to 3.12 c/bu when choice of test is allowed.

Effect of Testing Protocols

Buyers choose the tolerance which in turn defines testing protocols. In the base case a 1% tolerance was assumed. It is well recognised that tolerance tightening has the impact of raising costs and prospectively raising risks of not conforming. There are three elements of costs that are critical in evaluating effects of differing tolerance limits. These include testing costs, risk of not conforming, and adventitious commingling in the system. The first two are clear. As tolerance is tightened, testing costs increase and risk of rejections increase.

However, it is not known how adventitious commingling in handling (either through residual grains remaining in handling equipment or through cross contamination of lots) would be impacted by increasing/decreasing tolerance levels. To approximate for this, here give two cases. One has a tolerance level of 0.5% (50% of the base case value) in which here increased the levels of adventitious commingling that would be identified at this tighter specification so parameters of the distribution were twice that in the base case.

A second case was developed in which the tolerance level was assumed to be 5%. In this case, the parameters for the distributions for adventitious commingling were assumed to be 50% of base case levels. These cases illustrate the potential effect of increasing tolerances. In both cases, alternative tests which achieve these desired tolerance levels and their associated accuracies and costs were utilised rather than the parameters for the base case test accuracies and costs. For the 0.5% tolerance level, a strip type test was assumed having a cost of $40/test and accuracy of 99%. For the 5% tolerance level model, choice of tests was between two strip type tests costing $20/test with one having accuracy of 95% and the other 99%.

Increasing the tolerance further (0.5%) results in testing of every 5th unit, but changes from the country elevator when loading to the export elevator when receiving. The rejection rate at the importer increased as the tolerance became tighter. Rejection for a 5% tolerance was 1.07%, while at a 0.5% tolerance was 3.00%.

Costs and risk premiums also increased as tolerances tightened. Tightening the tolerance from 5% to 0.5% increased testing and rejection costs from 0.63 c/bu to 1.67 c/bu for all bushels and from 0.83 c/bu to 2.60 c/bu for Non-GM bushels. The risk premium increased from 0.47 c/bu with a 5% tolerance to 1.06 c/bu with a 0.5% tolerance. Total costs for Non-GM bushels increased from 1.45 c/bu with the 5% tolerance to 4.25 c/bu with a 0.5% tolerance.

14

Ethical Aspects of Crop Biotechnology

The Indian Patents Act of 1970 was amended to provide for product patents for agro-chemicals, drugs and pharmaceuticals and microorganisms over a transitory period. A product patent regime would be in place in India by January 1, 2005. The new patent law extends protection term for inventions from 14 to 20 years. To provide for protection of Intellectual Property Rights Regimes (IPRs) in respect of new plant varieties, Parliament has enacted a sui-generis legislation in the shape of the Plant Varieties Protection and Farmers Rights (PVFR) Act in 2001.

By prioritising FDI investments in biotechnology, the Government of India has sent appropriate signals to international investors. Further the fact that these changes have been in consonance with the WTO-TRIPs has lent greater creditability to these changes. Meanwhile, the Government has also brought in regulations over agricultural biotechnology. Apart from the biosafety regulations that form part of the Environment (Protection) Act 1986, the Indian Parliament has legislated the National Biodiversity Act in the year 2002, to provide for regulatory controls over access to biological resources in India.

Article 27(3) of the TRIPs Agreement lays down that members shall provide for the protection of plant varieties either by patents or by an effective sui-generis system or by any combination thereof. Article 27(3) in particular is critical to India as this requires us to provide protection of plant varieties by conferring plant breeders rights either by way of patents or through suigeneris plant variety protection legislations. Countries with strong R&D base in plant genetic engineering such as USA have robust Plant Utility Patent Legislations.

India is certainly not inclined to adopt patent protection regimes for its plant varieties. India is rather inclined to adopt a suigeneris legislation which is non-'patent' based. There are many reasons for this proclivity. India is one of the ten mega-diversity countries of the world and a rich storehouse of landraces of principal agricultural crops. India has a strong R&D base in conventional plant breeding methods. Its strength in plant genetic engineering is impressive but not an overwhelming factor by comparison. The first two strengths explain India's disinclination towards plant patent regimes or towards a sui-generis legislation which is 'patent'-driven.

While Plant Utility Patents Act provide for broad patents over plant varieties, traits and genes and even the physical parts of the plants, plant breeders rights provide for IPR only over varieties. As is well known, since 1990s the UPOV (Union for the Protection of Varieties) has been largely viewed by developing countries as offering the best regime for positioning their national legislations.

The central feature of the UPOV is the protection it affords to plant breeders who produce plant varieties that fulfil the criteria of distinctiveness uniformity and stability (DUS). The current version of the UPOV, viz. UPOV 1991 had added additional criteria of 'new' to DUS thus rendering DUS as NDUS. Contrary to the popular notion, the NDUS criteria of UPOV 1991 is not substantively different from the principles of 'novelty', 'inventiveness' and 'industrial application' (NII) which applies for patents. The criteria of 'novelty' and 'inventiveness' in Patent Laws are covered by the criteria of 'new' and 'distinct' in UPOV 1991. Thus by distinctness, the UPOV means a variety of plant which is 'clearly distinguishable from other varieties whose existence is a matter of common knowledge'.

It is apparent that this term captures the attributes of 'novelty' and 'inventiveness' implicit in Patent Laws. Even in respect of 'uniformity' and 'stability' criteria the UPOV does not offer different recipes. True, by 'stability', the UPOV conveys that 'relevant characteristics of protected plant variety remain unchanged either for a specified period or after repeated propagations or cycles of propagations'. It is also true that 'stability' is a difficult criteria for a plant breeder to fulfil. Attainment of 'stability' criterion is problematic for crosspollinated plants and non-single homozygous lines of autogamous plant varieties. This, in turn, reduces the commercialisation potential of the plant variety since inconsistency of genetic quality jeopardises commercial application of the plant variety. The same holds true of the criterion of "uniformity".

Therefore, the NDUS criteria of UPOV 1991 are homologous to the NII criteria implicit in Plant Utility Patent Laws. The UPOV has undergone two major amendments since its inception in 1961, i.e. in 1978 and 1991. The main concern with regard to currently operational UPOV 1991 has been its stringent provisions, which include the additional attribute of 'new' to the DUS criteria. The scope of UPOV 1991 in terms of coverage is larger. While plant varieties of all taxa are covered by UPOV 1991 only plant varieties of nationally designated species were covered by UPOV 1961 and 1978. Further by defining clearly that 'variety' represents a 'group of plants', or 'single' or 'several plants' or 'one' or 'several parts' of a plant(s) the UPOV 1991 has considerably enlarged the scope of its coverage.

Going by this definition even sexual and vegetatively propagated varieties are covered by UPOV 1991. The US PVPA legislation has been quick to adopt the UPOV 1991 formula. The amendments of 1994 to the US, PVPA broadens the term 'plant varieties' to include all materials harvested from the protected varieties. The other drastic feature of UPOV 1991 has been abridgement of the 'plant breeders exemptions' and 'farmers privileges' conferred by the earlier versions. Earlier versions of UPOV allowed farmers to re-use seeds generated from their farms for self-use, and allowed breeders to freely use protected varieties for further improvements. These concessions gave UPOV a 'less stringent' face in comparison to 'patent' laws.

UPOV 1991 gives an option to National Governments to disallow the farmers privileges of retaining or re-using seeds for self cultivation. UPOV 1991 also restricts the breeders rights by enlarging the right of the breeder to close variations of his cultivar or 'essentially derived varieties'. In other words, the breeder of a cosmetically bred variety would have to buy genetic dependency rights from the derived variety prior to commercialising the derived version. Further the duration of protection have also been extended under UPOV 1991 to 20 years for crops and to 25 years for trees and vines. Thus in terms of these rigorous features, UPOV 1991 eliminates the arbitrage between plant varieties protection and patent regimes.

Investments in Plant Biotechnology

The implicit rationale for IPR protection is that they promote investments in plant breeding and bio-engineering. Nevertheless it is a fact that the structure of IPRs do influence investments. IPRs induce their own pattern of innovations. The time taken for R&D to fructify as innovations is a crucial aspect guiding investments. Often IPR regimes play a key role in influencing the time pattern of innovations.

It is a well-known fact that conventional plant breeding methods are dilatory and time consuming. It requires 7 to 8 generations of repeated breeding to reduce heterozygosity of new genotypes. Arabica Coffee subject to three to four cycles of breeding still display heterozygosity and it is not until 20 to 25 years (spanning 4 to 5 generations) that one arrives at true breeding varieties. This time lag is even more pronounced for the Arabusta coffee variety (inter-specific hybrid of Arabica and Robusta Coffee) where segregation of unfavourable characteristics could still occur even after several generations of back crossing and selfing.

India released 306 rice varieties for planting during the period 1965-1986 after performing 20,000 crosses over a period of over 15 years since 1950. It must be understood that this long process of backcrossing was preceded by considerable time spent on location and identification of individuals plants with requisite traits. The speed or pace at which plant varieties qualify for being released for planting have economic implications on investment behaviour in seed/plant breeding industry.

Asexual modes of plant breeding which include plant genetic 'engineering' and transgenic techniques could obviate the long gestation period associated with prerelease preparations encountered conventional breeding. It is a different matter than biosafety concerns may delay releases of these varieties. The time lag in release of plant varieties fulfilling the criteria of novelty, distinctness, uniformity and stability (or alternatively the requirements of novelty, inventiveness and commercial application) raise fundamental issues from the view-point of capital investments. The time lag in varietal release is potentially least in the case of transgenic and non-sexually produced varieties while for traditional or classical methods the time lag can be greater.

The UPOV 1991 based plant IPR regimes could induce longer time lags as compared to the situation when an IPR regime is based on the 1978 version of the UPOV. This is due to the fact that the minimum genetic distance concept implicit in UPOV 1991 precludes 'close cousin' varieties from acquiring protection rights, unless the techniques of genetic distance determination by themselves are inadequate.

UPOV 1978 would facilitate capital investments to some extent as the release of plant varieties is periodically more frequent. Even under the plant patent regimes, investments could bring about short-term returns, though the volume of investments will have to be larger on account of capital intensity of advanced plant genetic engineering technologies. By comparison, UPOV 1991 regime is least conducive to capital investments as plant variety releases can be painfully slow under this regime. Therefore sustenance of conventional plant breeding under UPOV 1991 regimes, can only be ensured by infusion of low discount capital. Paradoxically it is the reduced availability of low discount public funds for plant breeding that has induced countries like India to go in for Plant Varieties Protection Legislations modelled on UPOV lines.

The implicit assumption in undertaking this step is that private funds for R&D in plant breeding will 'crowd in', if protection plant varieties rights are accorded. But the reality is that in liberalised deregulated developing economies, private investments have 'crowded in', in short return promising avenues. In other words, private investment funds for R&D are not likely to be channellised into medium to long return yielding areas.

Let r_a, r_b & r_c be the discount rates relevant for regimes A, B & C in Figure 1. Then the relative rates of discount will be

$$r_1 > r_b < r_c \tag{1}$$

and relative present worth will be

$$p_a < p_b > p_c \tag{2}$$

Consequently classical plant breeding activities will suffer for want of funds in the UPOV 1991. UPOV 1991 modelled national legislations will not necessarily hold attraction for private sector investors in R&D projects involving classical plant breeding as compared to say, UPOV 1978 or the Plant Patent regimes. Rather private investments would be catalysed better by sui-generis plant protection legislations, which partake of plant patent elements and focus in the main on hybrids and genetically engineered varieties.

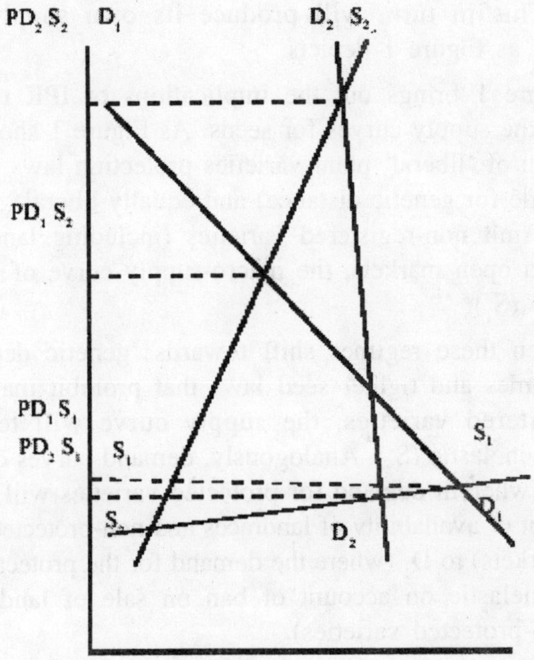

Figure 1: Supply-demand dynamics for seeds in different IPR regimes

Given the linkages between plant breeding and the seed industry these perverse investment behaviour trends can produce major economic consequences. The Seed Laws in various countries, both advanced and developing, are getting to be progressively oriented towards restricting market circulation to varieties that are conferred with plant breeder or patent rights. The UPOV guidelines on DUS are used world wide not only as the básis for establishing varietal distinctiveness and descriptions but also for seed certification purposes.

The DUS guidelines are employed for recognising and registering not only the 'basic' seeds used for multiplication of 'pedigree' seeds but also for certification of the pedigree seeds themselves. How absence of IPR rights will reduce Marginal Variable Costs (MVC) of seed multiplication. Tougher controls by certification of seeds could ensure that seed supply is kept in a 'constricted mode' and MVC of seed multiplication are raised. This in turn, will produce its own supply-demand dynamics as Figure 1 depicts.

Figure 1 brings out the implications of IPR regimes in terms of the supply curves for seeds. As Figure 1 shows that in a situation of 'liberal' plant varieties protection laws (which do not provide for genetic distance) and equally liberal 'seed laws' which permit non-registered varieties (including landraces) to be sold in open markets, the macro-supply curve of seeds will be elastic (S_1).

When these regimes shift towards 'genetic dependency' PBR regimes and tighter seed laws that prohibit marketing of non-registered varieties, the supply curve will tend to be relatively inelastic (S_2). Analogously, demand curves could shift from D_1 (wherein demand for protected varieties will be elastic on account of availability of landraces and non-protected varieties in the markets) to D_2 (where the demand for the protected variety will be inelastic on account of ban on sale of landraces and other non-protected varieties).

In the D_1–S_1 situation, supply constraints and demand flexibility could induce low price regimes while in D_2–S_2 the reverse situation would occur. D_2–S_2 situation is not desirable for countries like India for other reasons as well. The experience of transitional economies of East Europe, which had re-modelled their Seed Laws in conformity with European Community Seed Laws, has been sad. Farmers of the East European bloc cultivating and conserving landraces have been adversely affected by new seed laws, which by prohibiting trade in non-registered varieties not fulfilling the DUS criteria have contributed to the erosion of landraces in these countries. It is thus clear that agro-biodiversity can be adversely affected by the uniformity criteria practices. Herbicide-tolerant cotton has been adopted rapidly since its introduction in 1997.

In the USA, 80% of growers are making fewer tillage passes and 75% are leaving more crop residue. In a farmer survey, 71% of the growers responded that herbicidetolerant cotton had the greatest impact on soil fertility related to the adoption of reduced tillage or no-till practices. In soybean, the growers of glyphosate-tolerant soybean plant a higher percentage of their acreage using no-till or reduced tillage practices than growers of conventional soybeans. 58% of gyphosate-tolerant soybean users reported making fewer tillage passes than they did five years ago compared with only 20% of non-glyphosatetolerant soybean users. 54% of growers cited the introduction of glyphosate-tolerant soybeans as the factor that had the greatest impact toward the adoption of reduced tillage or no-till methods.

Negative Impacts of Bt Toxins

The use of GM crops can positively impact agricultural species biodiversity if the GM crops enable the management of weeds and insect pests in a more specific way than chemical herbicides and pesticides. In particular, the adoption of insect-resistant Bt crops, expressing highly specific Bt proteins, represents an opportunity to replace broad-spectrum insecticide use. The insecticidal proteins expressed in Bt crops such as Bt maize and Bt cotton are so narrow in their activity that they have little or no activity against non-target organisms. Furthermore, the toxins are expressed within the plant tissues, minimising the exposure of animals that do not feed on the crop plants.

As a consequence, considering the large number of field studies that have been conducted, few or no differences have been seen with respect to community structure or individual species abundances where fields of Bt crops have been compared with conventional crops that have not been treated with insecticides. Where they have been calculated, indices of species diversity and community structure have not differed significantly for Bt corn fields compared with untreated conventional corn fields or for Bt cotton fields compared with conventional cotton fields. The only species that have been observed to be significantly and consistently less abundant in fields of Bt crops compared with conventional fields are the target pests and their specific parasites.

In studies where the conventional crop fields have been sprayed for the target pest species of the Bt crop (as it routinely occurs in most crop systems) many nontarget species have been observed to be adversely impacted, leading to significantly lower non-target populations in sprayed conventional fields compared with Bt crop fields. With corn fields, this is particularly obvious for foliagedwelling species because of the method of application of these insecticides, but ground-dwelling species like carabids and cursorial spiders are also often affected, directly or indirectly, by insecticidal sprays and are apparently not affected by Bt corn.

Similarly, a variety of studies of Bt cotton in the USA, Australia and China have all demonstrated that populations of many non-target species are higher in Bt cotton fields than in sprayed conventional cotton fields. Likewise, work on potato fields in the northeastern USA has revealed larger populations of many generalist predators in Bt potato fields than in conventional potato fields treated with appropriate broad-spectrum insecticides. In contrast to Newleaf potatoes and microbial Bt formulations, however, the broad-spectrum insecticide, permethrin, had much broader and more severe unintended impacts on non-target arthropods.

Monarch Larvae in Bt Cornfields

The controversy surrounding the fate of Monarch butter- fly larvae in US Bt cornfields seems to be solved. The Bt protein built in transgenic corn resulted in toxic effects to the Monarch larvae, and triggered a worldwide protest against GM crops. Later, extensive field work demonstrated no significant impact of the Bt protein on Monarch larvae. Morerealistic conditions did not yield the same results, and in fact the lacewing larvae remained healthy. This discrepancy by the fact that prey fed to the larvae was fully vital in contrast to prey affected by Bt toxin.

British Farm-scale Experiments

The results of the three-year-long experiment on three genetically modified herbicide-tolerant (GMHT) crops over > 200 fields in Great Britain have had a great impact in the press and therefore on the public. The well intended experiments yield significant amounts of data related to herbicide and crop management differences and these were rigorously collected and duly peer reviewed.

The results can be summarised as follows: differences in biodiversity between crops exceed differences between GMHT and conventional crops. There were higher early season weed numbers and biomass in all three GMHT crops and higher weed mortality in GMHT sugar beet and canola resulting in lower lateseason biomass and seed rain of weeds in those crops, but lower weed mortality in GM maize. More detritivores (collembola) were observed in all three GMHT crops as a result of higher weed detritus.

There were lower numbers of bees, butterflies and Heteroptera in GMHT sugar beet and canola were observed as a result of reduced weed populations; generally higher numbers of invertebrates in GM maize. Lower herbicide inputs in GMHT crops. It has been argued that GM maize is performing better because it has been treated with the broad-band herbicide atrazine.

The GMHTcrops have been planted in Great Britain for the first time and farmers actually have not been experienced enough to apply advanced techniques such as no tillage, which would have then given the full advantages of the method. It is quite logical and has never been contested that the application of a broad band herbicide such as Roundup Ready® would be efficient in killing weeds, and as a consequence the biodiversity within the fields is reduced and has various consequences. The farm-scale studies could be summed up in a simple message: no weeds→no insects and→no weed seed. In turn, no insects and no weed seed→no bird food. No bird food→no birds.

However, it is not this simple; first of all we must realise that we are not dealing with natural habitats and even the sky larch is an artificial product of agriculture. We are therefore dealing with highly dynamic ecosystems and have many opportunities for improvement. Relatively minor change could bring back biodiversity to the fields by applying the appropriate methods. The Farm Scale Experiments (FSE) fail to take into account that management methods have changed in the USA with the advent of GM crops. Experimental outlays in field research must take into account the full potential of management in modern farming, such as no tillage methods.

Even seen as a true management experiment it is not done in a full farm scale manner: it fails to compare yield and other input-output data to residue analysis of conventional herbicides within the non-GM crop fields. It would have been possible to apply standard methods used in integrated pesticide management systems.

When, in the USA, large areas of crops were replaced by GMHT varieties, the profile of agrochemical inputs on the farm changed, the proportion of the land that was tilled before sowing sometimes decreased, less chemicals were lost in leachates and run-off from the field, and, as glyphosate and glufosinateammonium are relatively short-lived and of low toxicity to animals, the change in profile was considered to lessen the wider impact of farming.

The chainof impacts was not the samefor all crop species, and generalisations are difficult. A caveat to acknowledge is that the soils in the UK are not similar to those in the USA and this might reduce the potential benefits of no tillage strategies, a problem that should be tackled on an experimental basis. If all the FSE data were available and a better-adapted management had been applied, results would not look so bleak for the Roundup Ready® technology. Data from Romania have shown that economically it is indeed rewarding to use the Roundup Ready® technology.

Overall, with the flexibility and simplicity of the herbicidetolerant crop method it should be easier to make progress (which has its limits there, where farmers do not like weed components in the harvest because there are several problematic toxicity cases known to be connected to certain species of weeds. With the incentive of the economic advantage farmers will agree more easily to do something extra for agricultural biodiversity to enhance conservation in arable fields.

Statisticians should have a closer look at variation, dynamics and individual treatments. Some of those treatments could well reveal key data on how to enhance successfully biodiversity in fields with GMHT crops. Having a closer look at variation related to the individual management methods would probably also have the potential to suggest future strategies, but this exploration might be limited by the high amount of noise and variability in the system.

13

Risks of Segregating GM Crops

Development of genetically modified (GM) crops is challenging the functions and operations of the grain marketing system. The adoption of GM corn and soybeans in the United States has resulted in numerous interventions to ease the transition to marketing of these crops. The path taken in the case of GM wheats is more elongated for numerous reasons. In contrast to the other grains and oilseeds, commercialisation of GM wheats is evolving concurrent with a fairly extended process of public scrutiny and commercial concerns.

One of the more important concerns is that of testing and segregation. Given there will no doubt be market segments adverse to GM content in wheat shipments, adoption and efficient marketing of GM wheat will require protocols for contractual limits, testing and segregation. Implicit in these insinuations are that some buyers, for varying reasons including regulations and product marketing, may choose or have no recourse but to limit the content of GM wheat in Non-GM wheat purchases. Presumably, these buyers would do so by specifying in their purchase contracts some limit on GM content and/or more precise prescriptions regarding production/marketing/handling processes.

This is what has evolved in the commercialisation of other GM crops. At least initially, or indefinitely, one could envision a marketplace of buyers with differentiated demands for their aversion to GM content. Hence, it is critical to have a prescribed system that conforms to these requirements. Within the microstructure and economics of the grain marketing system, some of the important concerns with respect to GM crops marketing center on added costs and risks.

Additional testing involves added costs of conducting the tests, of which there are several technologies and varying accuracies. The risk is that of GM wheat varieties being commingled and detected in customers' shipments who place limits on GM content. This is indeed an economic problem as agents seek to determine the optimal strategy for testing and other risk mitigation strategies.

The primary focus is on testing and tolerance strategies confronting the U.S. marketing system, producers, processors and foreign processors. Here provides a quantitative model that can be used to assess costs and risk of alternative strategies for marketing GM crops. The distribution of costs and risks in the case of GM wheat have come to be an important prerequisite to further commercialisation of this trait. Most important, provide estimates of the risk premium necessary for suppliers to expose themselves to tolerances associated with Non-GM shipments. Though the problem is focused on wheat, the methodologies would be applicable to other crops, characteristics (e.g., vomitoxin) and production processes.

GM Wheats

Development of GM wheats has lagged other grains and oilseeds for varying reasons. Most important is likely the more complex genetics. Other contributing factors include: 1) wheat is a smaller volume crop within North America; 2) exports are of greater relative importance; 3) import country regulations vary much more for wheat and are less well-defined; and 4) competition amongst exporting countries is likely more intense and compounded by radically different marketing systems regarding quality and trade practices, etc.

These points notwithstanding, there are several initiatives for the development of GM wheats. In North America these have been primarily on the Round-up Ready® wheat trait, though there is extensive research elsewhere on a wide range of GM traits in wheat. Virtually all development in North America is currently on Hard Red Spring (HRS) wheats. Experimental trials are being planted in South Dakota, North Dakota, Minnesota (and no doubt elsewhere), as well as in selected Canadian prairie provinces.

Round-up Ready® wheat (RRW) is an example of 1st stage benefits. Other 1st generation benefits should be commercially available by 2005 (Bloomer). However, 2nd and 3rd generation effects will not be accessible until 2006 and beyond. In the case of wheat, 2nd generation effects would likely include enhanced protein quality, novel starch types (functionality), enhanced nutritional content, reduced allergens and improved freshness and shelf-life for baked products. These observations were echoed by Biane indicating that consumer benefits in the case of wheat include extending shelf life, improved nutrition, and reduced allergens.

The pressures for adopting GM wheat, specifically RRW, come from a combination of cost reduction, reduced dockage, increased profitability of competing crops (being recipients of GM technology), and the prospect of 2nd and 3rd phase benefits associated with GM wheats. In all cases, there is an insinuation that development of a GM wheat is good and should be pursued. However, it is very clear that GM wheats should not be commercialised, in whatever fashion, until some type of IP and testing system is developed to mitigate risks to buyers. The asynchronous regulations, along with selected buyer resistance and indigenous differentiated demands, ultimately suggest that a dual marketing system (or a marketing system to facilitate coexistence) is inevitable.

This is likely true in the domestic market even though labeling would be voluntary with different approaches likely adopted by buyers for branded versus non-branded (e.g., private label, food service, etc.) products. This would also occur internationally between countries with and without tolerance limits, and/or other requirements for the traits, and those with approved traits.

Finally, inevitably, tolerances will need to be defined and/or those proposed will be needing refinement. There are two forms in which tolerances are applied. One would be those defined by regulatory agencies. Second, would be as commercial tolerances. Most important in establishing these tolerances are that costs increase as tolerances are tightened, and that risks are mitigated by the use of tolerances. Risks are defined as buyers receiving a product that should be rejected and sellers having a product rejected that should have been accepted. There is a fundamental trade-off between risks and costs. Tighter tolerances result in increased costs and decreased risks.

Dual Marketing Systems of GM Wheats

Ultimately, an alternative to a regulated system is a system with dual market channels. Such a system is represented in Figure 1. All the basic elements are included from grower delivery, handling at country elevators (Sample and Test Rec., No-Test Receiving, Non-GM, GM, Non-GM, GM, Sample and Test Loadout, No-Test Loadout) and export elevators (Export, GM, GM, GM, Non-GM, Rec. Sample and Test, Non-GM, Load Sample and Test), and the potential for testing at each of these functions.

Thus, the system only involves movement to the point of first processing. An important and nontraditional practice reflected here is that growers declare GM content at point of first delivery. That is, subject to their own uncertainty about the GM content, growers would declare (i.e., as in some type of contract or affidavit) whether the content of the grain includes GM varieties. This is commonly referred as GM declaration and has been an important element of the evolution of the market of GM grains.

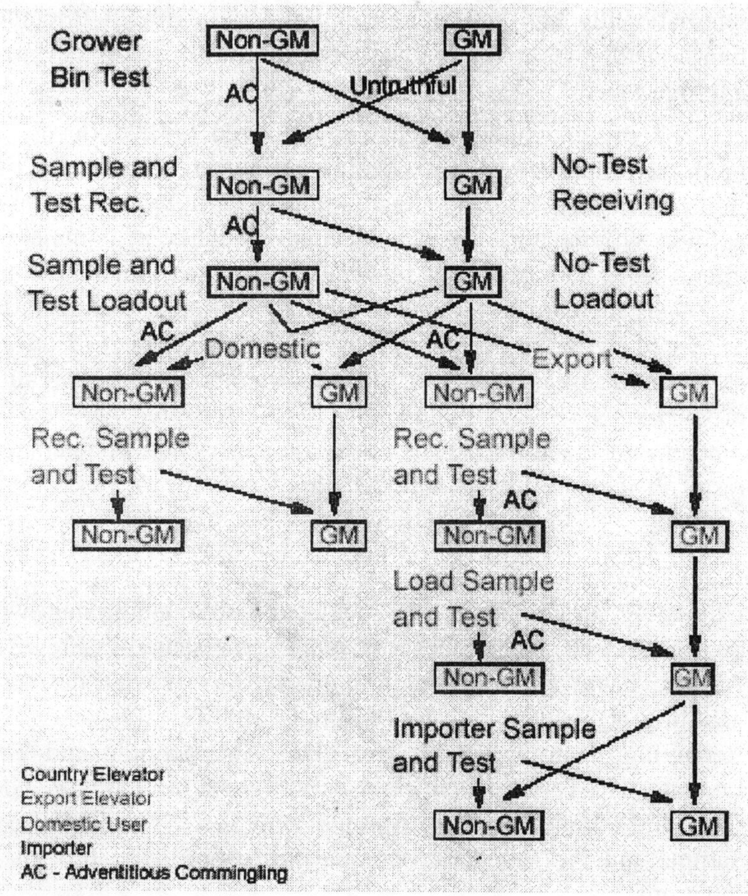

Figure 1. Grain Handling Subject to Adventitious Commingling.

This system could be envisioned as being adopted with several different scopes. It could reflect an elevator that seeks to segregate within their own facilities, or it could be elevators specialised in handling GM versus Non-GM. Or, it could be a vertically integrated firm with some elevators specialising in GM versus Non- GM handling. Each type of adoption has occurred in the marketing of other GM grains.

Risks are incurred throughout this system. There are three sources of grower risk. These include volunteers in subsequent crops, pollen drift, and on-farm adventitious commingling. Experience with volunteers has been limited in these crops for obvious reasons. Current literature suggests the level of risk of volunteers to be in the area of 31% of fields infested with an average density of 9 plants/sq. meter in the first year. The percent of fields infested and densities decline as years since the last wheat crop increase. By year 5, only 9% of fields were infested with an average density of less than 1 plant/sq. meter.

These results indicate that there is a positive incidence, and this declines through time and is dependent on variety and agronomic practices. Using reasonable assumptions about planting rates etc., these risks translate to a probability of about.009 in year 1 (which would apply if wheat were planted on ground that was planted to wheat in the prior year), and diminishes to virtually nil in the years following.

Pollen drift, in the case of self-pollinated GM wheats, is relatively modest compared to cross pollinated crops like corn. The rate of outcrossing is generally less than 1% but can range as high as 5% with pollen drifting from 5 to 48 meters. While handlers routinely segregate and blend grains as a primary function of their business, there is added risk of handling GM grains due to the possibility of adventitious commingling. If running elevators non-stop, contamination is 4%; after 3 minutes, it declines to.2%. These are corroborated by Hurburgh who suggested the sources of adventitious commingling at the elevator/handling function to be: handling .3%, shipping .3%, and mixing 1% for a total of =1.6% or a probability of .016.

Another source of risks is testing. Throughout the system there are risks associated with testing. Tests are not 100% accurate. However, the level of risk can be determined and varies with technology and tolerance. These are described below. Finally, inevitably a contract penalty may be imposed by the buyer if GM content is found in a Non-GM shipment. This may be a simple penalty, or a rejection of the shipment by the seller. In either case, costs to the seller would be accrued.

Testing and Tolerance

There are several aspects of testing that are important. Most important is that testing would only apply to Non-GM shipments. It would be unnecessary to conduct tests on those shipments/lots already known to be GM. Thus, testing would only occur for those shipments/lots that are thought to be Non-GM.

There are two basic tests that could be used for analysing for the presence of RRW. These are commonly referred to as strip tests and Polymerise Chain Reaction (PCR) tests. These tests are for "single-trait" events. The PCR test is based on DNA technology and is more commonly used in international contracts. Strip tests are or would be more commonly used domestically.

The grain marketing system is evolving and beginning to adopt these protocols in GM corn and soybeans. Of importance in each case, as it would be in wheat, are tolerances, testing technologies, frequency at which tests are applied, declaration of GM content at the country elevator, and associated added costs and risks.

End-users and buyers express their needs and aversion to GM in contracts with tolerances. This is critical. Ultimately, it is incumbent on those buyers wanting to limit GM content in Non- GM shipments, for whatever reason (commercial or regulatory), to specify limits/restrictions in their purchase contracts. Those not averse to GM would not have to do anything special.

This can be implemented in existing contract forms, and in a way similar to other factor limits. Specifically, for non-grade determining factors (e.g., dockage, vomitoxin, etc.) buyers specify limits in their contracts. This could be similarly accomplished for GM content. For example, a buyer may specify a limit simply as: not to exceed X% GM content and/or a discount may apply if the tolerance is exceeded. In addition, an acceptable test/sampling procedure would have to be concurred. Presumably, that would be standardised in such a way to make the contract language and implementation common across transactions.

Growers declare varieties (i.e., whether the shipment contains GM varieties) at time of delivery (Harl). It is important that the grower knows the variety being delivered or at least has the capability of knowing. This provides a wealth of information that needs to be conveyed to the marketing system. Not only does this provide the essential information for segregation and testing requirements, it has several other positive benefits.

Empirical Model

A model of grain flows reflecting the structure of a dual system with testing and segregation of GM/Non-GM flows (Figure 1) from growers to either importers (Importer, Sample and Test, Non-GM GM) or domestic end-users (Domestic, Non-GM, Rec. Sample and Test, Non-GM, GM, GM) was developed. The model assumes adventitious commingling can occur at various stages of the grain marketing chain with given probability distributions.

A level of GM/Non-GM adoption by farmers is assumed and farmers may/may not identify grain lots delivered as GM/Non-GM with a probability of "truth-telling." Tests are conducted at various stages to determine if grain indicated as Non-GM contains levels of GM exceeding tolerances. Non-GM flows exceeding the tolerance are diverted to GM flows at the stage of the marketing chain where they are identified and subjected to a penalty.

An important and innovative feature of the analysis relates to the risks the handler/shipper is exposed to and the consequence of violating a tolerance. For example, if a ship is being loaded with Non- GM wheat, and even though the shipper is taking grain from a segregated Non-GM flow, it is possible that the ship may be found to have a detected level of GM content (for example, BT or RRW corn materials in wheat). In practice this would be interpreted as a contract violation and subject to either rejection, penalty, or renegotiation, all at a loss to the shipper.

Any of these would be terms of the purchase agreement. In any case, the shipper would be subject to an implicit cost or "risk premium" associated with this type of content. The value of this risk premium (π) as the expected costs for a Non-GM system (EV_{NGM}) less the certainty equivalent (CE) of the utility of additional costs of a system containing both GM and Non-GM segregations and include it in cost function. This premium reflects the point at which decision makers would be indifferent to the current Non-GM system or a system handling both GM and Non-GM segregations.

The model is developed as a stochastic optimisation model of a grain marketing chain. The objective function contains a von-Neuman- Morgenstern type utility function, with decreasing absolute risk aversion and increasing relative risk aversion. The model chooses the optimal testing strategy (where to test and how often to test) that maximises utility by minimising additional system costs for a supply chain handling a portfolio of segregations representing two states of nature (GM and Non-GM grains). The portfolio utility is comprised of the weighted disutility of additional system costs for handling both GM and Non-GM segregations.

The model estimates the additional system costs due to testing and segregation for each of the segregations (states of nature) separately. Additional system costs are impacted by:

— C_{NGM} is additional testing and segregation costs added to Non-GM shipments to maintain GM separation,

— C_{GM} is additional costs for GM bushels (assumed zero),

— k is location in the system where tests can be applied (country elevator receiving, local elevator loading, export elevator receiving, export elevator loading, importer receiving, domestic user receiving),

— T_k is cost of individual test applied at location k,

— S_k is sampling intensity at location k,

— V_{NGMk} is volume (number of lots) of Non- GM handled at location k,

— D_k is discount or penalty applied to grain diverted from Non-GM to GM flows at location k, and

— V_{DGMk} is bushels diverted from Non-GM to GM flows at location k.

The model derives additional system costs at each stage of the marketing chain, tracks segregation flows throughout the system, and derives statistical properties on the proportion of shipments with GM exceeding specifications within end-use flows.

The model incorporates risk in a number of random variables. These include farmer "truth-telling;" adventitious commingling which occurs at several locations (farm, country elevator, export elevator, and transportation equipment) due to various factors (inadequate cleaning, etc.); sampling and inspection plans; and test accuracy.

The distribution of grower risks (inclusive of volunteers, pollen drift, and on-farm handling) were derived to reflect the risks depicted. Similarly, handling risks were taken to depict those reflective in Hurburgh and Casada, Ingles, and Maghirang. The penalty for GM contained in a Non- GM shipment was assumed to be uniformly distributed within a range of 40-90 c/bu in the export market and 2-20 c/bu in the domestic market.

There are several aspects of the cost components. First, it is a result of a contract specification agreed between buyer and sellers. Second, it is important whether the test is evaluated at origin (i.e., export port) or destination (import port). If the former, being out of contract is not as great. Finally, some export elevators (e.g., with shipping bins) may be more capable of testing prior to loading than others. The logic to the export penalties is based on two components. Discounts for GM in Non-GM corn have been in the area of 10% of the value, which in the case of wheat would be about 40 c/bu.

However, in some cases, rejection may entail reshipping the grain to some other market at a cost to the shipper. In many geographical locations internationally, this would be about the equivalent of 50 c/bu. For the domestic market, these would reflect handling costs and possible shipment to alternate destinations.

Simulations and Sensitivities

Simulations and sensitivities are then evaluated relative to this base case. Sensitivities were conducted to examine affects of risk attitudes, tolerance, variety declaration, level of GM adoption, level of discounts for rejection of Non-GM shipments, and choice of test type by location. A second model is developed to also examine impacts for shipments to domestic users.

Base Case

The base case was defined to reflect the most likely system and protocols. These include:

— Export shipment to importers;

— GM adoption by farmers of 20% (based on market distributions of GM aversion of buyers);

— Grower declaration of GM content at the country elevator;

— Testing was allowed at any or all of the following: Country Elevator (CE) at receiving and/or loadout and at the Export Elevator (EE) at receiving and loadout;

— Testing technology at the export/import level.

The optimal testing strategies which maximise utility (minimise disutility) of GM/Non-GM system versus the current Non-GM system. The optimal strategy would be to test every 5th railcar

of DUS, which brings about resultant changes in demand and supply position of seeds.

POST R&D REGULATIONS

Post-R&D regulations play a critical role in deciding on the pace of investments in the biotechnology sector. Even if a start-up enterprise enjoys a favourable IPR regime, that rewards its inventions, it is unlikely that it will invest in product development, in case post-R&D regulations are rigid and dilatory in nature. This then forms the backdrop for strategic alliances between start-up ventures and downstream industrial complexes. Multi-layered regulations in the post-R&D phase involve complexities which are expressed in the delays in commercial application /utilisation of invented products or technologies.

In the case of third generation plant biotechnologies, such as transgenics, the problem is compounded by the fact that regulations by themselves are complex and less understood by the regulators themselves. A transgenic plant would have to undergo multistage regulatory checks and clearances before finding commercial application.

India is uniquely situated as far as biotechnology investments and infrastructure are concerned. In the area of plant biotechnology, advances have been made in transgenics of rice and wheat carrying stress tolerant genes such as CodA, COR47 and HVA1. The larger challenge is to take these products through the regulatory processes. More formidable is the issue of market development and appropriate pricing of transgenic seeds. There are many imponderables here.

Market shares for seeds in India have varied from crop to crop and from region to region. The 'demand' for transgenic plants or seeds may be even more uncertain, given the 'safety' dimensions and risk aversion propensity of Indian farmers. In the dry agricultural pockets of Central and South India, farmers tend to be attached to traditional drought-hardy varieties of plants, which in turn contributes to their initial 'inertia' to shift to transgenics. Thus the task of marketing and pricing of novel biotechnology plants gets to be complex in such environment of 'demand incoherence'. A high margin pricing strategy for seeds may choke demand for transgenics.

Thus the degree of risks in commercialising transgenic plants could be of a high order. Given these realities, the idea of strategic alliances between start-ups and downstream entities looks to be realistic. The fact that premier R&D institutions such as the Indian Institute of Science, has set up its own start-up biotechnology venture, renders a strategy of tie-ups, a logical option. Much will depend upon the complexities inherent in product development and the willingness of competent downstream entities to signup agreements. To a large extent, this will depend upon enabling policies at the upper end of the value chain. To sum up, India's agricultural biotechnology industry has the necessary policy and regulatory support to promote active R&D.

However, in the critical area of product development and commercialisation much more attention needs to be paid. The multi-tiered regulatory framework that exists could seriously affect the performance of start-up companies that would like to run through the entire life-cycle of biotechnology product development. Consequently, partnerships and strategic alliances should be encouraged between start-up ventures and established companies that have high product development capabilities. The experience of other countries in the Asia Pacific region could be a key input in framing development strategies for biotechnology products in India.

15

FUTURE OF TRANSGENIC CROPS

Genetically modified crops created by modern agricultural biotechnology have attracted worldwide attention in the past decade. Cautious voices warn that the health and environmental effects of GM crops are uncertain and that their cultivation could have unintended adverse consequences. Alternatively, supporters of the technology assert that GM crops could revolutionise world agriculture, particularly in developing countries, in ways that would substantially reduce malnutrition, improve food security, and increase rural income, and in some cases even reduce environmental pollutants.

Can the GM crop movement develop into an agricultural revolution on the scale of the Green Revolution? To answer this question, first, it is important to consider what an agricultural revolution entails. Viewed historically, movements that come to be considered agricultural revolutions share the following features:

— The movements gave farmers incentives to produce—i.e., the technologies provided a net benefit to farmers.

— The movements substantially improved agricultural production, food nutrition, or both; or they substantially decreased necessary inputs such as fertiliser or water.

— People were generally willing to adapt culturally and economically to the new technologies, and consumers accepted the products of the agricultural movement.

— There was cooperation among those that provided the technologies, regulated the technologies, and used the technologies.

— The movements were sustainable, eventually without public subsidisation.

On a regional scale, GM crops might indeed be considered revolutionary—that is, they could meet all five criteria for an agricultural revolution. For example, in the United States, Canada, China, and Argentina, genetically modified varieties of soybeans, corn, and cotton now make up from about a third to 80 percent of total plantings of those crops, and provide benefits for growers such that these GM varieties will likely continue to make up a substantial portion of total plantings in the foreseeable future.

Likewise, policymakers and the general public in these nations are accepting of this new technology. Adoption of these GM crops has led to improved yield, decreased use of pesticides or particularly harmful herbicides, and, in some cases, improved food quality. While farmers in other nations, such as India and South Africa, have more recently begun to plant GM crops and experience the beginnings of a potential Gene Revolution, the revolution has yet to occur on a global scale. It has stalled because consumer and environmental concerns, along with precautionary regulations, have limited its spreading to the countries that could benefit from it most, notably much of sub-Saharan Africa where famine continually threatens the population.

The Green Revolution that had its origins in the 1940s, and reached its peak in the 1970s, continues to affect agricultural practices today. By analysing the Green Revolution's objectives, science and technology, sources of financing, regulatory environment, and ultimate successes and failures, an assessment of the ongoing GM crop movement—whether and how it might make a revolutionary impact on world agriculture. The stated objective of the Green Revolution was to increase food production in regions of the world facing impending massive malnutrition.

In the post-World War II era, scientists and policymakers considered those regions to be Latin America and Asia. Some argue, in retrospect, that this geographic choice was also motivated by Cold War politics: a largely U.S.-supported effort to prevent the spread of communism by ensuring adequate food supplies in at-risk countries. Regardless of its motivation, the introduction of high-yield varieties (HYVs) of crop seed, along with pesticides, fertilisers, and irrigation systems, transformed agriculture on those two continents.

With initial funding from the Rockefeller Foundation, individuals including U.S. plant breeders, agronomists, entomologists, soil scientists, and engineers worked in developing nations while training local agricultural scientists to extend the work in their own locales. The World Bank, Food and Agricultural Organisation of the United Nations (FAO), United States Agency for International Development (USAID), and other national and international organisations later joined the Rockefeller Foundation to make this effort succeed. And succeed it did, in terms of increasing food production in Asia, Latin America, and even parts of the industrialised world such as Great Britain.

In Africa, however, where the movement came later, the Green Revolution has yet to improve food production in a sustainable way. As such, this movement provides several important lessons for understanding the possible course of the Gene Revolution.

Green Revolution and the GM Crop Movement

Here, the Green Revolution and the current GM crop movement compared in four basic areas: science and technology, funding sources, where the movement occurred or is occurring, and the policies and political motivations surrounding each movement.

Science and Technology

The Green Revolution presented a considerable advance in agricultural technologies for farmers in the developing world, and, to a limited extent, in industrialised countries as well. For the first time, scientists and plant breeders integrated their research with farming practices in traditional agriculture to tackle problems that were constraining crop yield. High-yield seeds for rice, wheat, and corn were introduced in parts of the world where these crops made up a significant portion of the daily diet, and subsequently of food exports.

Pesticides, chemical fertilisers, and irrigation systems were also introduced to aid farmers in controlling previously unmanageable pests, dealing with low-quality soil, and delivering water to crops according to their requirements. The Gene Revolution, propelled by genetic engineering, allows previously unheard-of combinations of traits across species to achieve pre-specified objectives. For example, daffodil and bacterial genes can be introduced into the rice genome so that the rice produces beta-carotene, the precursor of vitamin A.

The benefits of the current varieties of GM crops include yield increase, reduced agricultural inputs such as pesticides and fertilisers, reduced vulnerability to the whims of nature, and improved nutritional content. For the most part, these benefits have been limited to parts of the industrialised world to which current GM crop development and marketing have been targeted and, among those, to countries that have allowed their cultivation. Other GM crops are now being developed that survive on less water, that survive in soil heavy in salt or metals such as aluminum, that convert or "fix" nitrogen from the air, and that produce vaccines against common diseases such as cholera and hepatitis B.

A fundamental challenge in this newest agricultural movement that did not arise during the Green Revolution is the definition and treatment of intellectual property (IP). IP issues are central to the Gene Revolution because whereas science and technology move forward through the sharing of ideas and resources, IP ambiguities and restrictions can often limit the valuable diffusion of science and technology. Commercial application of biotechnology has taken place primarily in the United States and primarily through the private sector.

The issue of who "owns" a particular event (the successful transformation) of a genetically modified crop and who can develop it further has become so economically important and contentious that numerous cases involving this issue are being litigated. Some observers consider IP issues to be among the most important impediments to the development and adoption of GM crops in the developing world. Patent rights that universities may have on their sponsored research, corporate profit interests, and the ability of farmers to buy IPprotected seed are salient IP issues.

Funding Sources

Philanthropic organisations, i.e., the Rockefeller and Ford Foundations, provided the backbone of early funding for the Green Revolution. The scientists who created high-yield seeds and their associated pesticides and fertilisers worked in conjunction with, and were funded by, these foundations along with the governments of Mexico, India, and several other countries. In 1971, while the Green Revolution was bearing its first fruits in many parts of the world, the Consultative Group on International Agricultural Research (CGIAR)—a system of 16 Future Harvest Centers working in more than 100 countries-was created.

With the creation of CGIAR, support for developing world agriculture became more broad-based and included European nations, Canada, and Japan. Genetically modified crops are largely the product of private industry. This is partly because new technologies are far more costly than existing ones, and the biotechnology industry was able to gather the necessary funds to develop these technologies long before public awareness of

GM crops could lead to publicly generated funding for GM crop development. Successful companies typically focus on their markets with the intent of generating profit.

With regard to agricultural biotechnology, companies in the United States and elsewhere have thus far created primarily seeds that farmers in industrialised countries can and will purchase: corn and soybeans that can tolerate a particular herbicide, corn and cotton that are resistant to particular pests, and food crops that last longer on the supermarket shelf. Because of the "technology fee" that growers pay to use these crop seeds (including recoupment of industry's research and development costs as well as profit), and because the seeds are designed particularly for their planting situations, the targeted farmers in industrial countries have generally found it worthwhile to buy these seeds and have been willing to pay the technology fee.

Thus, in industrialised nations, GM crop technology has had the potential to revolutionise farming. However, the current GM crop seed varieties are neither affordable nor useful to most of the poorer farmers in the world; hence, their revolutionary impact in the developing world has been limited thus far. Indeed, there seems to be a mismatch of setting and technology, due to the funding sources of basic research.

Some agricultural biotechnology companies have recently expressed interest in working with regional research institutions to develop crops that would be profitable and affordable for farmers in developing countries. In addition, they are willing to donate a substantial portion of their scientific knowledge, such as genomes of key food crops, to increase agricultural knowledge in the developing world. In this way, the challenges related to IP may be lessened.

The Green Revolution was a success, in terms of its stated objectives, in Mexico and the rest of Latin America, India, and much of Southeast Asia. On the other hand, the Green Revolution has had little significant impact in most areas of Africa. Two prominent hypotheses for this outcome are that the technology package that was so useful in some parts of the world was not applicable to African farms, and that rural transportation systems are ill-designed to deliver either the technologies or their resulting products.

The technologies introduced in Asia and Latin America in the Green Revolution generally required not more land, but chemical fertiliser and well-timed water. Farmers who could access these inputs did well while others did not. To the extent that large landholders also had access to fertiliser and irrigation, they tended to adopt the new technologies early and successfully.

It may be too early to predict the varying adoption rates and benefits of yet undeveloped Gene Revolution technologies given the differing characteristics of farmers and regions. What can be said from the Green Revolution experience is that farmers will not adopt and utilise technologies over the long term that do not cost considerably less than current technologies, produce considerably more than current technologies, or substantially reduce the variability of cost or production in their own locales.

As opposed to the Green Revolution, the key component of the Gene Revolution technology is improved seed. This being the case, all farmers, small or large, should be able to take advantage of the Gene Revolution; theoretically, the Gene Revolution is scale-neutral, providing that one can pay for the seed. However, cultural factors may deter farmers from embracing the new science; genetically modified crops have already become a stigmatised technology in some parts of the world because of concerns about manipulating organisms in seemingly "unnatural" ways and fears of unintended adverse impacts on the environment or human health.

Policies and Politics

At the time the Green Revolution was first seriously considered, the United States and the rest of the developed world feared that food crises in developing countries would cause political instability that could push those countries over to the Communist side. Partly as a result of this issue, the U.S. government was highly concerned about agricultural science in the developing world and worked with foundations and scientists in the post-World War II decades to bring about the Green Revolution in regions subject to famine.

As of yet, there does not appear to be a strong political motivation for genetically modified crops to succeed in the developing world. Communism is no longer a threat, and famines, while still a problem in parts of the world, appear to be more the result of localised weather, politics, and war conditions than a sweeping threat that commands sustained government and public attention in industrial countries. Instead, public concerns and national and international regulations are now the driving force behind whether GM crops are adopted or rejected in various parts of the world, because wider public scrutiny and the newness of the science have led to concerns about environmental and health risks of GM crops that must be dealt with at the policy level.

The battle between U.S. and European Union regulations, which feature very different stances on the acceptance of GM crops in food and feed, has been the major determinant of this outcome. In addition, a variety of nongovernmental organisations (NGOs) that are concerned about the influence of multinational corporations, environmental degradation, crop diversification, food safety, globalisation, and the influence of U.S. interests are prominent and influential in both the industrialised and developing countries. These NGOs were not nearly as influential during the Green Revolution.

The Gene Revolution far resembles the Green Revolution in the following ways:

— It employs new science and technology to create crop seeds that can significantly outperform the types of seeds that preceded it;

— the impact of the new seed technologies can be critically important to developing world agriculture; and

— for a variety of reasons, these technologies have not yet reached the parts of the world where they could be most beneficial.

On the other hand, the Gene Revolution is unlike the Green Revolution in the following ways:

— The science and technology required to create GM crop seeds are far more complicated than the science and technology used to create Green Revolution agricultural advancements;

— GM seeds are created largely through private enterprises rather than through public-sector efforts; and

— the political climate in which agricultural science can influence the world by introducing innovations has changed dramatically since the Green Revolution.

The similarities and differences between the Green and Gene Revolutions lead us to speculate that for the GM crop movement to have the sort of impact that would constitute an agricultural revolution, the following goals still need to be met and the related challenges overcome.

— Agricultural biotechnology must be tailored toward, and made affordable to, developing-world farmers. Unless these conditions are met, farmers may not see that it is in their best interest to use GM crops at all despite the unique benefits those crops could provide.

— There is a need for larger investments in research in the public sector. Numerous studies have shown the importance of public-sector research and development to aiding agricultural advancements, including the Green Revolution. Partnerships between the public and private sectors can result in more efficient production of GM crops that are useful to the developing world and can expand the accessibility of those crops and their associated technologies to developing-world farmers.

— To garner the level of public interest that can sustain an agricultural revolution, agricultural development must once again be regarded as being critically important from a policy perspective in both donor and recipient nations. As population numbers continue to increase today, agricultural development is more necessary than ever to eliminate malnutrition and prevent famine, particularly in sub-Saharan Africa. GM crops are seen by many as a means for addressing those problems. However, policymakers worldwide are far from being a combined force on this issue.

— Policymakers in the developing world must set regulatory standards that take into consideration the risks as well as the benefits of foods derived from GM crops. This goal is crucial to the cooperation of the many stakeholders that are affected by GM crops and also for the sustainability of the GM crop movement in the foreseeable future. Without regulations that explicitly take into account potential benefits to both farmers and consumers, those nations that might stand to benefit most from GM crops may be discouraged from allowing them to be planted.

Revised regulations on genetically modified crops must accompany widespread collective policy efforts to revitalise agricultural development. And before developing world farmers and consumers can benefit from GM crops or any other type of enhanced crop breeding, the technologies must be affordable and farmers must understand how to use them. The GM crop movement must overcome an intertwined collection of challenges before it can have an impact beyond those regions of the world that already produce excesses of food.

If the GM crop movement can overcome these challenges, while proving itself to be acceptably free of adverse health and environmental impacts, it has the potential to provide benefits to farmers and consumers around the globe in previously inconceivable ways, while mitigating the need to use potentially harmful chemicals or scarce water supplies for agriculture. It can then indeed become a true "Gene Revolution."

Bibliography

Altieri, M.A., "The ecological impacts of transgenic crops on agroecosystem health", *Ecosystem Health*, 2000.

______________, *Agroecology: The science of sustainable agriculture*, Westview Press, Boulder, 1996.

______________, *Biodiversity and Pest Management in Agroecosystems*, Haworth Press, New York. 1994.

Donegan, K.K. and R.J. Seidler, "Effects of transgenic plants on soil and plant microorganisms", *Recent Res. Devel. Microbiology*, 1999.

______________, Palm, C.J., V.J. Fieland, L.A. Porteous, L.M. Ganis, D.L. Scheller and R.J. Seidler, "Changes in levels, species, and DNA fingerprints of soil microorganisms associated with cotton expressing the Bacillus thuringiensis var. Kurstaki endotoxin," *Applied Soil Ecology*, 1995.

Duke, S.O., *Herbicide resistant crops: agricultural, environmental, economic regulatory, and technical aspects*, Lewis Publishers, Boca Raton, 1996.

Goldberg, R.J., "Environmental concerns with the development of herbicide-tolerant plants." *Weed Technology*, 1992.

Hilbeck, A., M. Baumgartner, P.M. Fried, and F. Bigler, "Effects of transgenic Bacillus thuringiensis corn fed prey on mortality and development time of immature Chysoperla carnea (Neuroptera: Chysopidae)", *Environmental Entomology* , 1998.

Kendall, H.W., R. Beachy, T. Eismer, F. Gould, R. Herdt, P.H. Ravon, J. Schell, and M.S. Swaminathan, "Bioengineering of crops", *Report of the World Bank Panel on transgenic crops*, World Bank, Washington, 1997.

Mellon, M and J. Rissler, "Now or never: serious plans to save a natural pest control." *Union of Concerned Scientists*, Washington DC, 1998.

Palm, C.J., D.L. Schaller, K.K. Donegan and R.J. Seidler, "Persistence in soil of transgenic plant produced Bacillus thurigiensis var. Kustaki endotoxin." *Canadian Journal of Microbiology,* 1996.

Paoletti, M.G., and D. Pimentel, "Genetic engineering in agriculture and the environment: Assessing risks and benefits," *BioScience*, 1996.

Pimentel, D., M.S. Hunter, J.A. LaGrow, R.A. Efroymson, J.C. Landers, F.T. Mervis, C.A. McCarthy and A.E. Boyd, "Benefits and risks of genetic engineering in agriculture." *BioScience*, 1989.

Rissler, J.; and M. Mellon, *The Ecological Risks of Engineered Crops*, MIT Press, Cambridge, 1996.

Saxena, D.S., Flores and G. Stotzky, "Insecticidal toxin in root exudates from Bt corn", *Nature*, 1999.

Snow, A.A., and P. Moran, "Commercialization of transgenic plants: Potential ecological risks." *BioScience*, 1997.

Index